THE END

OF

THOUGHT

Novels by Joseph P. Cody

THE TIGER'S FURY

DRAGON FANG

HUBOT - Human Robot

METANOIA - Total Conversion

PHANTOM TEAM

WILD VIOLETS - Growing Up
In The 1940s And 50s

FIND SPARTACUS

THE END OF OF THOUGHT

JOSEPH P. CODY

The views expressed herein are solely those of the author.

This book is written, printed, and bound in the United States of America

200417-1ST

ISBN: 978-0-9791167-9-7

A Publication of:

Autotech Industries
688 – 11th Avenue NW
St. Paul, Minnesota 55112

Autotech Industries is a publisher; it does not sell books. This and other books by Joseph P. Cody may be ordered from Amazon.com or any book store.

To Eric

And to a Society of Reason . . .
Where the Truth, the Good, and the
Beautiful Prevail

CONTENTS

PART I

EVOLUTION

PART II

THE ESSENCE OF TIME

PART III

LORDS OF THE WORLD

INTRODUCTION

The title of this book, *The End Of Thought*, can evoke two meanings. The first is where The End Of Thought means its purpose. This could lead one to think the book is about why people think. While a worthwhile endeavor, that is not the intended meaning. What is meant is the secession of thought as in there was a time when people thought and now they are moving rapidly to a point where they do not think. It will be the over arching premise of this tome to show that there has become less and less thought over the past century or two. We now find ourselves at a point where at best thinking is hardly considered worth doing and at worst it is looked on as unworthy of effort to down right silly and ultimately as dangerous to the individual and to society.

It will further be argued that the poison that has invaded the common psyche first started in the western world and more recently with the advent of electronic devices has spread over the entire planet. What is that poison? It is a number of ideas that started out in the rarified atmosphere inhabited by intellectuals in various fields. In time all of the new ideas of philosophy and science coalesced around one idea that enraptured the smart guys as well as all other strata of society. And that one idea is the strange thing called *evolution*. It seems everyone believes in the theory of evolution in some way—the small group of strict creationists being the only exception. And even there, one could not be too certain that in some way they have difficulty expunging evolution entirely from their thinking because of one thing or another such as the fossil record.

Why would anyone say the belief in evolution would cause the end of people thinking? To begin with it is important to note that there will be no attempt to say people can no longer think clearly enough to plan their day. However, when a whole world society succumbs to saying that they believe that the thing called evolution is a proven scientific fact, we are in trouble. In fact it is always approached from the point of view that they *believe* in the scientifically proven *theory* of evolution—an oxymoron if there was ever one. The concept of belief is never far from the subject of evolution. This is not the same, on the one hand, of the case of a religious belief, like the belief that Jesus rose from the dead, to use a commonly accepted one by Christians. On the other hand it is not the same as saying one believes the sun will rise tomorrow morning, either. We never use the word belief with such a concept—the sun *will* rise tomorrow. But, when one speaks of evolution both of the fore mentioned cases are present and that causes a type of cognitive dissonance. That is to say, there are facts like there are fossilized bones of extinct animals, and it is evident that there is variability among members of a living species. But, there is no proof connecting these to evolution.

If you do not believe what was just said, simply mention evolution to anyone from those with advanced degrees to common working people and you will hear some of the craziest, most illogical statements there ever were. One of the all time favorites is that the theory of evolution is every bit as proven as the law of gravity. If that really were the case why is one call a theory and the other a law? But, not to get bogged down, every second of every day everyone sees the law of gravity working. In any discussion of evolution you can mention that if one drops a pencil, there's proof of gravity. Such a statement will be taken without undue rancor. However, make another equally valid statement such as there is no recorded case of a new species ever coming into existance and from that point on logic and common sense are jettisoned.

The point of this is when an entire population buys into an absurdity things simply never work right. Religious belief is one thing. We all have what we call a belief system about life and more importantly what happens after life. That's fine. But when we all hold in common a contradiction in terms and logic, that is trouble.

This author has a friend with an advanced degree in a life science. He of course was all in favor of evolution. After a few letters exchanged with him on the subject I asked him to give me the title of a book that he had read that made the definitive case for evolution. His reply was that any elementary textbook on biology will nicely fill the bill. By saying that he indirectly said he had never read a single book on the subject, and yet he was firmly convinced there could be no valid argument that said it was not true. In other words, he, true to form, held evolution as an article of faith.

This speaks to an old saying to the effect that we are the most certain about the things we know the least about. For example, most people have a firm opinion about what the Unites States should do about the conflicts in the Middle East. They watch the news on TV and may read a newspaper article about them now and then. But when it comes to buying a health insurance plan it's different. In that case they study various plans and may even make a chart of the advantages and disadvantages of each. What this says is it is the details that complicate things. Thus it was with my friend. An elementary text on biology would contain nothing but a few platitudes about evolution. When confronted with numerous problems with evolution my friend was adrift because he had never had his thinking on the subject challenged, or more to the truth, he had never thought about it at all.

What follows will not be an attempt to discredit evolution, as such. However, there will be an attempt to show where it came from and in so doing one must necessarily recount various claims of the theory, who first presented them and how those claims have became so imbedded in the collective mind. It will also be necessary to point out where various claims go wrong and that may look like an effort to discredit the theory but that will not strictly be the case. Evolution is an ideology, and no amount of reason or fact will impugn something like that. However, in order to show the absurdly of evolution it will be necessary to take various tenets of that system of thought and take them apart. This is because there are a lot of well meaning people, many of them quite conservative, who have never thought much one way of the other about evolution.

Purpose

The purpose of this book will be to bring together what is known to humans concerning who and what we are without the blinders imposed by modern society on our thinking. It will include knowledge from several fields of study including science and philosophy. The intended readers will be people looking to learn something. That of necessity precludes those who have closed minds and that population unfortunately includes the majority of people. Most of what will be discussed could be found in any number of sources but there are some descriptions and conclusions that the reader may find new to him. Frequently it will happen that the reader will read a conclusion and say to himself, well, yes, of course. It is the case of "things you know but you do not know." What that means is we all have facts at our disposal on any particular topic, but until someone draws a rather obvious conclusion that those facts demand we are surprised that we had never thought of it. When we hear the conclusion we realize it had been before our eyes all the time and blurt out, "Of course, I know that," but in fact we did not know it. Without such conclusions one is not able to act intelligently on facts.

As a starting point it will be necessary to define philosophy and thereby indirectly science. That may seen odd because science as understood in today's world is the study of nature in so far as nature can be reduced to formulas and numbers and what does philosophy have to do with that? That will become evident when we examine a true definition of philosophy as follows: "philosophy is the science of all things naturally knowable to man's unaided powers, in so far as these things are studied in their deepest causes and reasons." [1]

What's going on here? We just said philosophy is a science. Many scientists today think that science and philosophy are mutually exclusive fields of study. However,

> *Science*, considered objectively, is a body of related data, set forth systematically, expressed with completeness and presented together with the evidence (proofs and explanations) which justifies and establishes those data as certain and true. *Science*, considered subjectively, is knowledge in the mind of a person; it is knowledge that is rounded, systematic, evidenced and complete. . . .

---

[1] Paul J. Glenn, *An Introduction To Philosophy*, St. Louis, Mo., B. Herder Book Company, 1943, p. 3.

When we say that philosophy is a science, we take the term *science* objectively. We mean that philosophy is a body of related data that is systematic, complete, evidenced and certain.

It is to be noted in passing that the evidence or proof requisite for science is not merely evidence from experiments or laboratory evidence. Evidence may also be (as in pure mathematics) *reasoned* or *rational* evidence. This point is important because many teachers of our times have presumed to limit science to the domain of the laboratory technician or the statistician arbitrarily ruling out rational evidence from the realm of true science. Such a ruling is . . . self-contradictory. For no amount of laboratory data, no number of experiments or catalogue of statistics, can amount to scientific evidence unless reason reduces them to unity and order and draws conclusions from them. And neither the nature and value of reasoning nor the basic force of the conclusions drawn by reason can be tested by laboratory devices or proven by experimental methods. We therefore reject the *positivistic, sensistic, materialistic* and *empiricist* doctrine that pure reasoning is of no scientific value. Philosophy is a rational or reasoned science, not a laboratory science. Philosophy does indeed use the findings of the laboratory sciences, but it is not confined or hampered by their limitations. It sheds great light on the data of the laboratory sciences. . . . But it is not fettered by their methods and subjected to their special requirements. Emphases in the original.[2]

What accounts for the disjunct between scientists and philosophy? The reasons are many but one will certainly be that one branch of philosophy is theology which in turn can include faith, religion, and spirituality. "Don't you dare bring any of those into my science," say many scientists. There it rests. They would not know the difference between theology and religion but do not bother them with splitting hairs, their minds are made up. The definition of philosophy above does not preclude studying things in their most basic state by applying numbers to observations where that is possible.

---

[2] Ibid. pp. 5-6.

A few additional words should be said about another part of the definition of philosophy. They are, "all things *naturally* knowable." What is meant by that is it precludes Divine Revelation. But, to anyone who wants to truly get to the bottom of things it is only common sense to use all the information available regardless of its source. Consider the following illustration. "Science" assumes time will go on forever. However, from Divine Revelation we know that at the end of time God will come to judge the living and the dead. And at the last judgment the present world will pass away and there will be a new heaven and an new earth; eternity will start. That means that there will still be people alive when time ends. That is something no amount of laboratory experimentation could discover, certainly not with the level of technology we have today. But, what if a theoretical physicist decided to examine all the measured data as well as the theories of the universe and time with the starting point that time ends when people end. He would not have to say where he got such a notion, only that he assumed it. Maybe, that would lead him to new discoveries no one had ever considered.

Two Truths

Since the first two sections of this work, the first on evolution and the second on time, are at least quasi-scientific matters and as just shown there should be no distinction between philosophy and science, a word must be said about a subject that causes no end or rancor between scientists and philosophers. That issue is called two truths or twofold truth. This says in essence that what is true in science need not be true in philosophy, faith, theology, religion or spirituality and visa versa. Some scientists do not have a problem with science versus philosophy but beyond that they draw the line. Since this idea is an important foundation stone of modern thought it will be necessary to review it at some length.

Twofold truths seems to have started with some Arabians chief among them was Averroes a Spanish-born Arab Moslem of the twelfth century. "This pernicious doctrine holds that what is true in philosophy may be false in theology, and visa versa." [3] Notice that the distinction here is

---

[3] Ibid., Glenn, p. 107.

between philosophy and theology. The term science did not exist in the middle ages and today science and philosophy, properly considered, are one and the same.

Glenn goes on to say, "The twofold-truth doctrine was taught in the 13th century by Peter d'Abano and John of Jandun in Italy, and by Siger of Brabant in the University of Paris. The doctrine is wholly indefensible, and it leads directly into the insane self-contradiction of skepticism. It is ruinous of all knowledge, of all science, of all philosophy." [4]

What's more, "The doctrine of twofold-truth is no longer defended by theorists . . . but it endures in practice, especially in the form of a twofold morality. Thus there are people who will justify sharp practices and open savagery in quoting as sound principles the silly cliches, 'Business is business' and 'All's fair in war,'—as though the business man and the soldier has a set of moral laws for office hours or term of service, and another set for private life." [5]

Another look at is as follows:

Thomas [Aquinas] was sent back to Paris in 1269 to challenge Siger of Brabant and his "two truths" doctrine. Siger claimed that rational truth and religion truth were two separate truths, which could contradict each other. The individual was to keep each in a separate compartment of this brain, as it were, calling upon each as needed and not worrying about any contradictions. The "two truths" was in utter opposition to all that Thomas Aquinas wrote and taught and he proceeded to demolish Siger of Brabant's ideas. God is the source of all truth; He is in fact infinite truth. God reveals religious truths to us directly, and He permits us to determine natural truths through natural reason. But, God is the source of both. Therefore, contradictions cannot possibly both be true. [6]

A final quote is from *The Catechism of the Catholic Church*, 1994

---

[4] Ibid.
[5] Ibid.
[6] Anne W. Carroll, *Christ the King Load of History*, 1994, Rockford, Illinois, Tan Books, pp. 186-7.

edition.

159 *Faith and science*: Though faith is above reason, there can never be any real discrepancy between faith and reason. Since the same God who reveals mysteries and infuses faith has bestowed the light of reason on the human mind, God cannot deny himself, nor can truth ever contradict truth. Consequently, methodical research in all branches of knowledge, provided it is carried out in a truly scientific manner and does not override moral laws, can never conflict with the faith, because the things of the world and the things of faith derive from the same God. The humble and persevering investigator of the secrets of nature is being led, as it were, by the hand of God in spite of himself, for it is God, the conserver of all things, who made them what they are.

This leads us to a few comments on what follows. As mentioned the first part will be concerned with examining evolution from several points of view, though hardly exhaustively. That will be followed by Part Two concerned with time. This is seen as important because evolution is nothing without time—lots of it. The first part of that section will be devoted to simply discussing several aspects of time because it is necessary for the reader to get used to thinking about a subject that is rarely brought to mind. Then will follow some work on what controls the flow of time, something even less the object of discussion and thought. After the discussion on time Part Three will argue the case of what or who is the driving force behind keeping the idea of evolution so solidly in place and what purpose there is to having the thought process of humanity collectively truncated.

PART I

EVOLUTION

Chapter 1

The Fathers of Evolution

There is probably no place in modern thought where people's ideas are more incoherent than on the subject of evolution. When it comes to scientists, they are either deeply conflicted or have given in and accept evolution as an article of faith all the while saying and, indeed, thinking it is an article of science. If they are among the former they never escape a level of mental agony when it comes to that subject. However, that mental state will allow them to have a reasoned discussion about evolution, at least at times. From the latter group one will be met with outright hostility for even mentioning the subject. That is all the more distressing because "evolution" is more than any other subject the keystone of modern thought and not in a positive way. Evolution sounds good in abstract and general statements, but as will be shown it does not fare well in the particular.

As already mentioned there will be no attempt to refute evolution because millions of people have spent billions of hours coming up with ways to "prove" it is true and have failed, and on the other side they have spent an equal effort trying to destroy any and all arguments against it. The hope here is that people who have never given the subject much thought will be interested enough to read at least sections of what follows. There are ideas that are repeated in different contexts and it is hoped that this will not put off the reader because they are so far from the normal discourse on evolution that they bear repetition. Now, we must start our task of meandering through the modern mind by an examination of the origins of evolution.

Where did Evolution Start?

Evolution is not, as many or most people think, a modern theory. Evolution and the arrival of distinct species was a subject of lively discussion by the ancient Greek philosophers, circa 500 BC. Add to that it did not spring fully developed in its present form from Charles Darwin, either. It had a long gestation period in Europe after what is called the middle ages. We will see here that it was not really a scientific theory at all. It was the product of the thinking of the time.

Therefore, we will begin our rather cursory tour of the history of evolution with the Greeks. If one goes on the Internet and looks for articles about the ancient Greek notion of evolution, he will find most of them will be written by modern evolutionists who miss the main points made by the Greeks and instead interject their own opinions on the matter. Here an effort will be made to be more rigorous in stating what they really thought.

Aristotle

A careful examination of Aristotle will show he first believed in substance, something not at all admitted by modern scientists. Substance is a unity of primary matter and form. That is, the two are totally united to produce a "thing." Primary matter is an obscure concept and is not understandable alone but only in relation to form. To be knowable, determinateness or form is required. When speaking about species Aristotle said,

> . . . the true method is to state what the definitive characteristics are that distinguish the animal as a whole; to explain what it is both in substance and in form [that is, its substantial from], and to deal after in the same fashion with its several organs, in fact, to proceed in exactly the same way as we should do, were we giving a complete description of a couch. [1]

Aristotle accepted teleology. Teleology is defined as the quality of being directed toward a definite end; natural phenomena determined by an overall purpose in nature. Evolutionists believing in the development of life by blind chance cannot, of course, accept that there is a goal or purpose to life.

Aristotle found teleology so evident in nature that he asked himself how his predecessors had been able to avoid seeing it there, or,

---

[1] Etienne Gilson, *From Aristotle To Darwin And Back Again*, Notre Dame, Indiana, University of Notre Dame Press, 1984, p. 16.

still worse, had denied its presence. He explained their error on the grounds that they were deceived on the notions of matter and substance. The subsequent history of philosophy ought to confirm the correctness of his diagnosis, for insofar as the Aristotelian notion of substance as the unity of matter and form survived, the notion of teleology remained indisputable; but as early as the seventeenth century Bacon and Descartes deny the notion of substantial form (a form which constitutes a substance by its union with a given [primary] matter) and the notion of final cause becomes inconceivable. In fact, substance defined by its form is the end product of generation. That which remained, once the form was excluded, was extended matter, or rather extension itself, which is the object of geometry and is susceptible only to purely mechanical modifications. Descartes consigned the entire domain of living beings, including the human body, to the realm of mechanism. [2]

From the above we see that as long as the idea of substance remained, teleology was safe. But, by jettisoning substance teleology failed as well and we were left with mechanists determining what people thought. This is where modern science went off the rails and started referring to living things as aggregates. See the section on Substance vs. Aggregate below.

The Modern Lineage of Evolution

It is difficult to decide where to start our tour of the development of evolution in more modern times. As noted above it could start with Bacon and Descartes, or as will be shown later in a different context we could start with Sir Isaac Newton. But, to develop the present notion of evolution we will stick with the direct lineage of that thought and start with Erasmus Darwin.

Erasmus Darwin (1731-1802)

In the era of our concern we can start with Erasmus Darwin who was Charles Darwin's grandfather. He is significant because he wrote the book *Zoonomia (Laws of Organic Life)* 2 vol. 1794, in which he set out a plan of transmutation based on competition and sexual selection. One of the reasons his book did not generate more interest was because he had

---

[2] Ibid., Gilson, p. 21.

no scientific data to back up his ideas. However, it does show that Charles Darwin did not generate his theories solely on his own research and mental deliberations.

James Hutton (1726-1797)

James Hutton is considered one of the founders of modern geology. He is credited with introducing the idea of uniformitarianism. That's something of a mouthful that means the climates and geological mechanisms that we see today on the earth have been uniform, that is, the way they are now, for as long as life has been around. It is one of the bedrock assumptions of evolution. He said it this way:

> Not only are no powers to be employed that are not natural to the globe, no action to be admitted of except those of which we know in principle, and no extra-ordinary events to be alleged in order to explain a common appearance, the powers of nature are not to be employed in order to destroy the very object of those powers; we are not to make nature act in violation to that order which we actually observe, and in subversion of that end which is to be perceived in the system of created things. In whatever manner, therefore, we are to employ the great agents, fire and water, for producing those things which appear, it ought to be in such a way as is consistent with the propagation of plants and the life of animals upon the surface of the earth. Chaos and confusion are not to be introduced into the order of nature, because certain things appear in our partial views as being in some disorder. Nor are we to proceed in feigning causes, when those seem insufficient which occur in our experience. [3]

What is striking about that statement is that he is imposing his will on the way nature must act. ". . . the powers of nature are not to be employed in order to destroy the very object of those powers. . . ." That can only mean that he is demanding that chance can only operate one way. A chance combination of molecules in a direction that favors life cannot be undone by a reverse chance. That is, "Chaos and confusion are not to be introduced into the order of nature. . . ." Today it is almost universally

---

[3] James Hutton, "Theory of the Earth," *Transaction of the Royal Society*, Edinburgh, 1785, Vol. I, p. 301.

accepted that "chaos" did happen once approximately 66 million years ago when a large meteor hit the earth near the Yucatan Peninsula forming the Chicxulub Crater which is credited with causing the dinosaur extinction. The geological record is replete with other, though less dramatic, "chaotic" events.

Notice, Hutton is not talking about evolution specifically as evidenced by his choice of words, ". . . system of created things." Geology was his business though he saw that the progression in the forces at work in the earth's crust corresponded to the apparent changes of living things. It is also certain that the idea of evolution was part of his thinking. The word "evolution" had not come into use yet but "progressive development" and "transmutation" were used. However, a creator was not allowed to operate in the development of things ". . . and no extra-ordinary events to be alleged in order to explain a common appearance. . . ." That is, no use of intelligence—something that is beyond nature—is permitted. That made him a deist, one who believes that God created the universe and its laws but then took no further part or interest in its functioning.

As can be seen, by uniformitarianism they mean primarily to reject the possibility of supra-natural, that is intelligent, forces being necessary to cause the appearance of species. All that is needed is nature bumping along doing what it does to accidentally produce life and all the assorted species. If it takes millions of years for a species to form, severe disruptions of the environments in which they gestate cannot be allowed.

Chevalier de Lamarck (1744-1829)

As mentioned above, today most people think Darwinism means that the stress of a changing environment causes changes in species. That particular idea was put forth by Chevalier de Lamarck. He certainly did not invent the idea. It was more that his writings pushed it to the fore at the critical point in Western history when thinking was changing. Yet, even among scientists today, it is likely that not one in a hundred you asked could make the distinction between Lamarck and Darwin. To quote Lamarck on the subject of stress causing changes in species:

> Every new need necessitating new activities for its satisfaction, requires the animal either to make more frequent use of some of its parts which it previously used less, and thus greatly to develop and enlarge them; or else to make use of entirely new parts, to which the needs have imperceptibly given birth by efforts of its inner

feeling . . . . All the acquisitions or losses wrought by nature on individuals, through the influence of the environment . . . all these are preserved by reproduction to the new individuals which arise, provided that the acquired modifications are common to both sexes, or at least to the individuals which produce the young. [4]

Therein lies one of evolution's main flaws. ". . . provided that the acquired modifications are common to both sexes. . . ." That says a positive mutation must be acquired by both a male and a female in the same generation and that those two individuals must find one another and mate and produce fertile offspring. The long shot of modifications (mutations) coming at the appropriate time gets even longer by those obvious constraints.

In commenting on Lamarck's development of new parts Etienne Gilson says the following.

. . . it remains the case that the organs are born, grow, and form themselves *in order* to satisfy the needs of the organism. That an organ should strengthen itself by exercise is comprehensible, and, in any case, it is observable; but that an organ should be born simply because a living body has need of it is a quasi-magical operation. It is nevertheless by means of 'observation' of this order that Lamarck claims 'to demonstrate' that the use of an organ, and the efforts made to use it in novel circumstance, not only reinforce and enlarge it but even create new ones [organs] to carry on functions that have become necessary.' How can we imagine the birth of a new organ as the effect of its exercise, since that which does not exist cannot be exercised? Emphasis in the original. [5]

Charles Lyell (1797-1875)

Charles Lyell was also a geologist who believed in uniformitarianism. He is significant to our present discussion mainly because it was his work that popularized this idea. In one of his letters concerning his upcoming book on geology *Principles of Geology*, 3 vol., 1830-33, he stated the following.

---

[4] Chevalier de Lamarck, *Philosophie zoologique*, as quoted by Gilson p. 53.
[5] Ibid., Gilson, pp. 53-54.

It will not pretend do give even an abstract of all that is known in geology, but it will endeavor to establish the *principle of reasoning* in the science; and all my geology will come in as illustration of my views of those principles, and as evidence strengthening the system necessarily arising out of the admission of such principles, which as you know are neither more or less than that *no causes whatever* have from the earliest time to which we can look back, to the present, ever acted but those *now acting*, and that they never acted with different degrees of energy from that which they now exert. [6] Emphasis in Clark.

Here we again see the clear notion that ". . . from the earliest time to which we can look back, to the present . . ." geological processes and the general environment of the planet are assumed to have remained unchanged.

Charles Darwin (1809-1882)

When one speaks of evolution the conversation nearly always turns to Charles Darwin. In 1859 Darwin published a book called *On the Origin of Species*. Here we must add a side note. Ask any evolutionist, in fact ask anyone at all, who is the "father" of evolution and they will say Charles Darwin. Ask them if Darwin ever wrote a book and they will answer, "Of course. It's call The Origin Of Species." But that is not true. The title of Darwin's book is, *On The Origin of Species by Means of Natural Selection; or, the Preservation of Favored Races in the Struggle for Life*. And, one can see why the title is normally abbreviated as we shall do, too, referring to Darwin's work as *On the Origin of Species*. Could you, the reader, have given its full name? It is doubtful that many scientists have read the book, in fact, few of them have ever seen a copy of it.

Over ten years (1859–1869) Darwin published six editions of this work. Interestingly, he does not use the word evolution once until the sixth edition where he uses it eight times in over 500 pages.

Darwin Finds a Workable Theory

Charles Darwin studied species intensely for years observing how

---

[6] Quoted by Robert T. Clark and James D. Bales in *Why Scientists Accept Evolution*, Grand Rapids, Michigan, Baker Book House, 1966, p. 16.

their various features, traits and habits fit with the etiological niches they occupied. The following quote taken from Darwin's autobiography is interesting. In it he mentions Malthus' book dealing with economic capitalism. As noted above Darwin published *The Origin of Species* in 1859 so he worked on it a long time after arriving at a ". . . theory by which to work."

> In October 1838, that is 15 months after I had begun my systematic inquiry, I happened to read for my amusement, "Malthus on Population." And being well prepared to appreciate the struggle for existence which everywhere goes on from long-continued observation of the habits of animals and plants, it once struck me that under these circumstances favorable variations would tend to be preserved and unfavorable ones would be destroyed. The result would be the formation of a new species. Here then I had at last got the theory by which to work. [7]

Darwin thus says he found the basic theory of natural selection not from nature but from an economics book, though, as shown above the basic ideas had been percolating around for some time. And Malthus' book was a reflection of the prevailing societal thought.

Darwin was a scientist and specifically a biologist. His work was on natural selection. He started by collecting data about specific things he observed and from them tried to form his facts into generalizations or hypotheses. Opposed to this, evolution was a philosophical idea that had been around from the time of Aristotle as we saw above. In his sixth edition of *On The Origin Of Species* Darwin even mentions Aristotle. The point here is that the idea of evolution was nothing new to the eighteenth and nineteenth centuries, though the word had something of a different meaning at that time than it has today.

The commonly seen pictorial representation of a slithery thing coming out of the water followed by something on four legs, then a short very stooped sort of a monkey thing, followed by a better monkey, then better yet until finally a fully erect man, naked of course—twitter, twitter through the second graders—is not Darwin. In fact he was indignant with and did a fair amount of battle against those who tried to associate the philosophical concept of *evolution* with his science of *natural selection*.

---

[7] *The Autobiography Of Charles Darwin from the Life and Letters Of Charles Darwin* edited by his son Francis Darwin, no date given, p. 33.

A few words must be said about the science and the philosophy as used in the paragraph above. The Greeks were philosophers. They observed the world around them and tried to draw logical conclusions about the universe and man from what they could see. Mainly the started from generalizations and tried to explain specific things using them, that is, they proceeded with deductive reasoning. For example, they had arrived at the conclusion there were four elements or kinds of atoms: rock, air, fire and water. Applying their atoms to gold they could see from its obvious color that gold had more fire atoms in it than, say, lead which led to the alchemist's dream of making gold out of lead by putting more fire atoms into it. Or, looking at cucumbers they inferred that as they ripened and became golden it was because more fire atoms from the sun had entered into them. That is, in their thinking they went from the general to the specific. In this line of reasoning evolutionists took evolution as a given generalization and attempted to apply it to specific cases. Opposed to this, Darwin started collecting observations and eventually came up with the general idea of natural selection, that is, he used inductive reasoning.

Herbert Spencer (1820-1903)

It was Herbert Spencer who popularized the idea and the term of evolution. It is rather obtuse how Spencer defined evolution, ". . . Evolution is always to be regarded as an integration of Matter and dissipation of Motion, which may be, and usually is, accompanied by other transformations of Matter and Motion." [8] He also defined evolution this way, "Already we have recognized the fact that the evolution of an organism is primarily the formation of an aggregate, by the continued incorporation of matter previously spread through a wider space." [9] Notice how he thinks species evolve by sweeping up matter that happened to be lying about.

This second definition is easier to understand, though it is totally at odds with reality. One must remember that the philosophers at the time of Darwin and Spencer were conversant with the philosophy of Aristotle and the Scholastics. As a result they knew very well—at least should have—what a substance was as opposed to an aggregate.

---

[8] Ibid., Gilson, p. 80.
[9] Ibid.

Thomas Henry Huxley (1825-1895)

Thomas Huxley is important to our story mainly because he was a good orator and erudite writer. Searching for a substitute for creationism he latched onto Darwin's ideas and was his most vocal supporter. His efforts more than those of any other individual, including Darwin's, were responsible for the eventual acceptance of natural selection and evolution in general. His son Bateson Huxley writes:

> Under the suggestive power of the *Origin of Species* all those scattered studies fell suddenly into due rank and order; the philosophic unity he had so long been seeking inspired his thought with tenfold vigor, and the battle at Oxford in defense of the new hypothesis first brought him before the public eye as one who not only had the courage of his convictions when attacked, but could, and more, would carry the war effectively into the enemy's country. And for the next ten years he was commonly identified with the championship of the most unpopular views of the time . . . . [10]

Lord Kelvin, President of the Royal Society, awarded Huxley the 1894 Darwin Medal and said this about Huxley's part in spreading Darwinism.

> To the world at large, perhaps, Mr. Huxley's share of moulding the thesis of Natural Selection is less well known that is his bold unwearied exposition and defense of it after it had been made public. And, indeed, a speculative trifler, revelling in the problems of the 'might have been' would find a congenial theme in the inquiry how soon what we now call 'Darwinism' would have met with the acceptance with which it has met, and gained the power which it has gained, had it not been for the brilliant advocacy with which in its early days it was expounded to all classes of men. [11]

Alfred Russell Wallace (1823-1913)

Wallace came to the theory of natural selection independently of Darwin, though, Darwin was the first to use the term. In 1858, the year

---

[10] *Life and Times of Thomas Henry Huxley, Leonard Huxley*, Vol. I, 1903, pp. 259-274.
[11] Ibid.. p. 301.

before the publication of *The Origin of Species*, Darwin and Wallace jointly published a paper titled "On The Tendency Of Species To Form Varieties; And On The Perpetuation Of Varieties And Species By Natural Means Of Selection" in the *Journal of the Proceedings of the Linnean Society of London, Zoology* Vol. 3, (20 August): pp. 45-50.

The ideas of Darwin and Wallace were much the same but not identical. Darwin started with analyzing variation under domestication and carried that thinking into the world of nature. That is, he believed nature, given more time, was nothing but a stock breeder. Wallace rejected domestic breeding as completely unrepresentative and artificial and that domestically bred varieties would in a few generations revert to the wild, natural stock from which then came. However, the fact that both men came to very similar conclusions shows the state of the research and thinking of the time.

It goes without saying that there were many more men who made significant contributions to the concept of evolution and particularly to popularizing it. But those mentioned above were the ones whose ideas made it "work" in the minds of those bent on making evolution into an accepted fact, a "reality."

Chapter 2

A Look at Darwin's Book

Why was Darwin the one who became the father of evolution? One important reason was timing. As just mentioned Wallace came up with the same ideas and published them. Also as mentioned, Huxley promoted Darwin and his ideas with an enthusiasm that was not natural to Darwin. However, in this author's view one of the reasons lies in the way *The Origin of Species* was written, though in many respects it was the way such books were written in that day.

To start with, *The Origin of Species* was an impressive, comprehensive work covering all aspects of how species could change starting with artificial breeding of domestic plants and animals. The first edition was 440 pages. He lists innumerable observations he had made in the areas of plants, insects, birds, and animals. This included barnacles and worms. He had eclectic interests—if it was living he studied it and included many of his observations in his book. The very broadness of his observations adds immeasurably to the believability of his conclusions.

On the other hand, *The Origin of Species* was not a scientific publication like we think of technical research papers today. It contains no tables of data, no charts, no graphs. All of his arguments are made in prose. As an example of many such cases one finds the following:

A part developed in any species in an extraordinary degree or manner, in comparison with the same part in allied species, tends to be highly variable. Several years ago I was much struck with a remark, nearly to the above effect, published by Mr. Waterhouse. I infer also from an observation made by Professor Owen, with respect to the length of the arms of the ourang-outang, that he has come to a nearly similar conclusion. It is hopeless to attempt to convince any one of the truth of this proposition without giving the long array of facts which I have collected, and which cannot possibly be here introduced. I can only state my conviction that it is a rule of high generality. I am aware of several causes of error, but I hope that I have made due allowance for them. (Emphasis in the original.) [1]

It is not my contention to say that if he had included "the long array of facts" that he had collected it would have shown the stated "rule of high generality" to be false. However, they are not there and he so much as says, "Trust me, I'm honest." It will also be noted that the truth or falsity of this particular conclusion would not make his hypothesis of natural selection rise or fall.

To get a feel for how *The Origins of Species* was written this author turned to Chapter 4 on Natural Selection since that is the premise of the whole work. As such one would expect the facts and arguments to be scrupulously selected and keenly drawn. The following was selected since it would naturally draw ones attention. It is taken from the section entitled, "Illustrations Of The Action Of Natural Selection."

In order to make it clear how, as *I believe*, natural selection acts, I must beg permission to give one or two imaginary illustrations. Let us take the case of a wolf, which preys on various animals, securing some by craft, some by strength, and some by fleetness; and let us suppose that the fleetest prey, a deer for instance, had from any change in the country increased in numbers, or that other prey had decreased in numbers, during that season of the year when the wolf is hardest pressed for food. I can under such circumstances see *no reason to doubt* that the swiftest and slimmest wolves would have the best chance of surviving, and so be preserved or selected, provided always that they retained strength to master their prey at this or at some other period of the year, when they might be compelled to prey on other animals. I can see no more reason to doubt this, than that man can improve the fleetness of his grey-

---

[1] Charles Darwin, M.A., *On The Origin of Species by Means of Natural Selection; or, the Preservation of Favored Races in the Struggle for Life,* London, John Murray, Albemarle Street, 1859, first edition, p. 136.

hounds by careful and methodical selection, or by that unconscious selection which results from each man trying to keep the best dogs without any thought of modifying the breed.

Even without any change in the proportional numbers of the animals on which our wolf preyed, a cub *might* be born with an innate tendency to pursue certain kinds of prey. Nor can this be thought very improbable; for we often observe great differences in the natural tendencies of our domestic animals; one cat, for instance, taking to catch rats, another mice; one cat, according to Mr. St. John, bringing home winged game, another hares or rabbits, and another hunting on marshy ground and almost nightly catching woodcocks or snipes. The tendency to catch rats rather than mice is known to be inherited. Now, if any slight innate change of habit or of structure benefited an individual wolf, it would have the best chance of surviving and of leaving offspring. Some of its young would *probably* inherit the same habits or structure, and by the repetition of this process, a new variety *might* be formed which would either supplant or coexist with the parent-form of wolf. Or, again, the wolves inhabiting a mountainous district, and those frequenting the lowlands, would naturally be forced to hunt different prey; and from the continued preservation of the individuals best fitted for the two sites, two varieties *might* slowly be formed. These varieties would cross and blend where they met; but to this subject of intercrossing we shall soon have to return. I may add, that, according to Mr. Pierce, there are two varieties of the wolf inhabiting the Catskill Mountains in the United States, one with a light greyhound-like form, which pursues deer, and the other more bulky, with shorter legs, which more frequently attacks the shepherd's flocks. [2] (The author has highlighted words and phrases of uncertainty).

An analysis of the above excerpt is in order. First, it must be said that those two paragraphs are representative of Darwin's writing style. He was a good writer; his work flows nicely and has the ability to convince the reader of the soundness of the proposition. A person reading his book would not likely stop and scrutinize that selection any more than most of the rest of the book. However, once chosen this author reread it several times while analyzing the science as opposed to the writing style. In the

---

[2] Ibid., pp. 86-87.

process several interesting things came to stand out.

As a first point one finds it curious that Darwin must revert to "imaginary illustrations" to present his thesis of natural selection rather than presenting actual cases. Beyond that, there are several conditional words and phrases as have been highlighted. Since it is imaginary nothing can be said about the actual content except for the parts ascribed to Mr. St. John and Mr. Pierce. Those passages are stated as facts, and as such, add elements of soundness to the whole illustration.

Concerning Mr. St. John's cats, it is obvious to anyone who has been around animals that individuals of a species vary the same as people do. That some cats prefer to hunt mice and some rats is in the same order of variability as some cats are black, some white, some calico none of which proves anything other than the obvious, certainly not the means by which new species appear.

Of more concern is the reference to the two types of wolves in the Catskill Mountains. It would have been nice if the actual measured lengths of the legs of the two types of wolves had been given as a comparison because, after all, this is a scientific work.

The real problem comes with the actual citation of those wolves at all. Certainly shepherds had not been grazing flocks of sheep in the Catskills for more than two hundred years prior to Darwin's book, and it is doubtful if it had been that long. He is assuming that the more bulky wolves with shorter legs had selected themselves out in that short time. This is to say nothing of the kindly shepherds who kept herds of sheep at the disposal of the preying wolves year round so the wolves would not starve to death. Take it from one who was raised on a farm, predators were hunted and trapped mercilessly.

It is possible the whole case of the wolves was like the peppered moth (see below) and both types of wolves as stated—including all varieties in between—simply represented the normal variability of the species. In that case the greyhound-like wolf would be just as likely as the shorter legged ones to attack sheep. No predator passes up an easy meal.

However, now intrigued but the Catskill wolves above mentioned, the author tracked down Mr. Pierce and where he published his scientific findings. They appeared in an article in *The American Journal of Science and Arts*, where there is a single paragraph concerning the wolves.

Of wolves, two varieties inhabit the Catskill mountains; one called by hunters the deer wolf, from its habit of perusing deer, for which his light gray hound form adapts him. The other of a more clumsy figure, with short legs and large body, more frequently depredates on flocks under the protection of man. Foxes and rabbits are numerous, the white hares, martins and hedgehogs sometimes seen, squirrels seldom. [3]

Since *The American Journal of Science and Arts* is by its title also about arts, the memoirs of Mr. Pierce might have been included in that journal under the category of art. It should be noted that the case of "natural selection" among the Catskill wolves that Darwin uses to bolster his thesis is nothing but hearsay information from local hunters—hardly what one would call scientific facts.

Not to belabor the point, but it appears at this point that Darwin is slipping into writing friction, and not good fiction at that. Any noteworthy work of fiction contains a plethora of actual facts about the setting of the story. In fact, a good fiction writer has actually been to the places where the story's action occurs. An example is *Heart Of Darkness* first published by Joseph Conrad in 1899. It is set on the Congo River in Equatorial Africa. From his descriptions one could assume that author had actually made a trip up the Congo River which, in fact, he had. That's what gives the story much of its intensity.

By the sixth edition Darwin backs off somewhat from the two varieties of wolf being a great deal different, but they were still included. In any case Darwin gives no reference as to where Mr. Pierce published his findings. It is of note that Darwin frequently mentions the work of others and never gives references as to where the reader can find the presumably published works to which he refers. However, since *The Origin Of Species* is such a pivotal work in the rise of the concept of evolution, others have hunted down the references so they are readily available. The issue of *The American Journal of Science and Arts* where Pierce's article was published is available at The Library Of Congress's website.

Associated with the aforesaid comments it is of interest that in the first edition Darwin uses the phrase "I believe" 89 times, "perhaps," 76 times, "it seems," 33 times, etc. The way it is written one hardly notices those words or phrases of uncertainty. Especially, they would be passed over if the reader were predisposed to believe what he was reading. It should also be added that the present author tends to prefer quoting the

---

[3] James Pierce, "A Memoir On The Catskill Mountains." *The American Journal of Science and Arts*, Vol. VI, 1823, pp. 86-97.

first edition of *The Origin of Species* because it is the volume that launched Darwinism.

It is fair to say that the above analysis of a few paragraphs of *The Origin Of Species* has left this author more than a little disappointed. One would have thought that a purportedly scientific work that has so totally changed people's thinking on the subject of were they came from would have been written with so little precision with regard to the facts. Other parts of the book vary, some more precise, some less.

This author has never seen a published work where the author really looked at the actual text of Darwin's book with an eye to what was indeed being claimed or assumed versus what was proven or demonstrated by facts. Lest the reader think the above example was taken out of context another will be offered. If this seems to be a case of taking Darwin to task, so be it.

The author wanted to identify Darwin's distinction between what he calls varieties and species since the whole subject of *The Origin of Species* is, of course, species. As a result the following series of passages are used for that purpose. Let us start with a passage from his summary of his work.

> On the view that species are only strongly marked and permanent varieties, and that each species first existed as a variety, we can see why it is that no line of demarcation can be drawn between species, commonly supposed to have been produced by special acts of creation, and varieties which are acknowledged to have been produced by secondary laws. [4]

He starts with "On the view . . ." which is nothing but an opinion. He soon comes to " . . . we can see . . . ." Really? Who is this "we" that can see that? It certainly does not include this author. And his statement concludes by stating ". . . which are acknowledged . . . " Once again one must ask, by whom? In a statement of sixty words he has three imprecise, personalized, phrases of uncertainty.

To continue. That portion of the summary refers to the main body of the book as follows.

> Laying aside the question of fertility and sterility, in all other

---

[4] Ibid., Darwin, p. 408.

respects there seems to be a general and close similarity in the off-spring of crossed species, and of crossed varieties. If we look at species as having been specially created, and at varieties as having been produced by secondary laws, this similarity would be an astonishing fact. But it harmonizes perfectly with the view that there is no essential distinction between species and varieties.[5]

By the sixth edition he has changed the "Laying aside the question of fertility and sterility. . . ." to "Independently of the question of fertility and sterility. . . ." It also seems clear that Darwin uses varieties to denote what are commonly called subspecies.

One hates to unduly criticize, but "laying aside the question of fertility and sterility" is not an option. Without fertility there would be no species, no life. And, the difference between varieties and species is precisely that members of different varieties—of the same species—can interbreed and produce fertile offspring while members of different species cannot. In order to have a scientific discussion it is necessary to have defined terms. If members of two distinct species can mate and produce fertile offspring one would have to say they are not strictly speaking distinct species. Darwin goes to some lengths to keep bringing in the relatively rare cases where members of what have been *defined* as different species can mate and produce fertile offspring, at least at times, without giving specific examples. This simply muddies the water as it seems it is frequently his intent.

And, exactly what does Darwin mean by a variety? His definition is as follows.

And what are varieties but groups of forms, unequally related to each other, and clustered round certain forms, that is, round their parent species? Undoubtedly there is one most important point of difference between varieties and species; namely, that the amount of difference between varieties, when compared with each other or with their parent species, is much less than that between the species of the same genus. But when we come to discuss the principle, as I call it, of Divergence of Character, we shall see how this may be explained, and how the lesser differences between varieties will tend to increase into the greater differences between species.[6]

---

[5] Ibid., p. 243.
[6] Ibid., pp. 57-58.

Once again he makes the statement, "Undoubtedly there is one most important point of difference between varieties and species . . . ." and does not mention the most important point of difference namely that two varieties of the same species can mate and produce fertile offspring and two species cannot.

Above Darwin mentions Divergence of Character. We see what he means by that in a section on the subject.

DIVERGENCE OF CHARACTER.

The principle, which I have designated by this term, is of high importance on my theory, and explains, as I believe, several important facts. . . . As has always been my practice, let us seek light on this head from our domestic productions. . . . We may suppose that at an early period one man preferred swifter horses; another stronger and more bulky horses. The early differences would be very slight; in the course of time, from the continued selection of swifter horses by some breeders, and of stronger ones by others, the differences would become greater, and would be noted as forming two sub-breeds; finally, after the lapse of centuries, the sub-breeds would become converted into two well-established and distinct breeds [does he mean species?]. As the differences slowly become greater, the inferior animals with intermediate characters, being neither very swift nor very strong, will have been neglected, and will have tended to disappear. Here, then, we see in man's productions the action of what may be called the principle of divergence, causing differences, at first barely appreciable, steadily to increase, and the breeds to diverge in character both from each other and from their common parent.

But how, it may be asked, can any analogous principle apply in nature? I believe it can and does apply most efficiently, from the simple circumstance that the more diversified the descendants from any one species (p. 105) become in structure, constitution, and habits, by so much will they be better enabled to seize on many and widely diversified places in the polity of nature, and so be enabled to increase in numbers. [7]

He is saying that the lesser differences between *varieties* will tend to

---

[7] Ibid., p. 103.

increase into the greater differences between *species*. It may be a nice thought, and he certainly has convinced himself of that, but as with most of his theory he offers no proof. And near the beginning of both paragraphs he used the phrase "I believe" in case the reader missed them. He has a way of writing in which such words somehow do not register in the reader's mind.

Why Was Darwin's Book so Pivotal?

To get to the root cause of why Darwin's book was so pivotal we must turn to Huxley once again.

> That which we were looking for, and could not find, was a hypothesis respecting the origin of known organic forms which assumed the operation of no causes but such as could be proved to be actually at work. We wanted, not to pin our faith to that or any other speculation, but to get hold of clear and definite conceptions which could be brought face to face with facts and have their validity tested. The 'Origin' provided us with the working hypothesis we sought. Moreover, it did the immense service of freeing us for ever from the dilemma—Refuse to accept the creation hypothesis, and what have you to propose that can be accepted by any cautious reasoner? In 1857 I had no answer ready, and I do not think that any one else had. A year later we reproached ourselves with dulness for being perplexed with such and inquiry. My reflection, when I first made myself master of the central idea of the 'Origin' was, 'How extremely stupid not to have thought of that!' I suppose that Columbus' companions said much the same when he made the egg to stand on end. The facts of variability, of the struggle of existence, of adaptation to conditions, were notorious enough; but none of us had suspected that the road to the heart of the species problem lay through them, until Darwin and Wallace dispelled the darkness and the beacon-fire of the 'Origin' guided the benighted. [8]

As he says, "The facts of variability, of the struggle of existence, of adaptation to conditions, were notorious enough," but it took Darwin to see that one might postulate from those facts a method by which species could come into existance. However, standing back and looking at the

------

[8] Ibid., Huxley, Vol. I, pp. 241-244.

situation under normal circumstances, most people would hardly arrive at the conclusion that variability, struggle and adaptation by themselves would be likely to produce significant changes in species let alone entirely new species, let alone all species. But, one must keep in mind that they were desperate in wanting to reject creation and would grasp at anything to replace it.

All of that having been said, it was a struggle to get the idea of natural selection and evolution in general accepted. *The Origin of Species* was not a run-away best seller. Huxley put his reputation on the line in support of the idea. None of its proponents would allow that the ideas of Darwin and Wallace were anything but an hypothesis that was severely wanting in hard evidence. But the publication did not escape the notice of the biologists and other scientists of the time. In the sixth edition, Darwin has an entire section devoted to answering objections of his critics.

Chapter 3

Species, Evolution and Natural Selection

Species have already been mentioned several times in different contexts particularly relative to attempting to determine the distinction between varieties and species as used by Darwin. Now, the subject will be discussed again relative to evolution and natural selection.

Species

At the end of the book by Etienne Gilson mentioned above, there is a twenty page Appendix II where the author comments specifically on Darwin's famous book. In it he makes the following observations. First of all, Darwin was not concerned about the origin of anything. He fastidiously ignored any attempt to find or explain origins. He simply started with existing species and then looked at existing body parts and how they were perfectly adapted for that species to fill its ecological niche. That is, he was concerned with the survival of the fittest species not where species came from. He then *assumed* that the living thing—plant or animal—had adapted to changing conditions through natural selection to get the way they were. That was all he did, that was all he studied, that was all he cared about.

The next thing is even stranger. Darwin had no idea of what a species was, as shown above, even though the word is used often in his book. Gilson says regarding Darwin's use of the word species, "He had need of a word precisely to be able to deny its existance." [1] To repeat, the old standby definition of a species is a group of living things that can reproduce

---

[1] Ibid., Gilson, p. 170.

and generate fertile offspring. A clear definition like that was never mentioned but will be the one used in this book. If evolution, indeed, all of biology, is to have any hope of being scientific there must come a point where firm definitions are required. We demand clear definitions in other areas of our lives. Imagine the difficulty of determining your power bill if no one could agree on the definition of a kilowatt-hour.

However, the reason for evolutionists not wanting a clear definition of species is obvious. Using an unambiguous definition as above, it is a scientific fact—one of the few facts concerning evolution—that no new species has ever been seen to come into existence. By using a variety of less precise, in fact, rather confusing and even contradictory definitions that nasty fact can be more easily overlooked.

Darwin discussed human intervention in changing species mostly in the form of plant hybrids but he also discusses stock breeding. Darwin even went so far as to say that horse breeders who bred strong horses for plowing or fast ones for racing actually made new species even though any of these could cross breed and produce fertile offspring.

To him nature was nothing more than an invisible horse breeder. But, there's the rub. Horse breeding requires intelligence—oops. But, not oops at all. Darwin granted the intelligence of the stock breeder. Then without a pause, he went on to say that nature, acting by blind chance alone, had a lot longer time so did an even better job. What all evolutionists accept as an article of faith is that chance produces better and better results which is totally contrary to fact and common sense. Chance works equally both ways as we will see with influenza a few pages hence. Choice works in one direction, the direction of the intelligence making the decisions. What makes that so awfully hard to understand and accept?

What makes it so difficult is evolutionists will admit that chance works both ways but that the bad changes are even worse for survival so individuals with negative changes will immediately be selected out to die. However, what's to say that in the next few generations the animal with a positive change will not have a negative change that cancels out the positive one resulting in the positive change being lost? The insistence that in the long run positive changes are never lost is unrealistic, that is, unscientific.

Underlying all of this is the unspoken, even unknown, idea that life must succeed. They feel in their guts that life must keep getting better and better without the slightest idea why that should be so. This speaks of telos which they deny.

An additional point about species is the people in Darwin's time felt species were virtually synonymous with creation. The educated class would have known about fossils and might have had an uneasy feeling that there was something unexplained there. Not so with the average person, that is, until the advent of universal education.

To the honest man, eliminating species left the problem of the variety of living things and the fact of life itself unanswered. The mechanistic view that life happened by random chances was not satisfying. This led to yet another idea that the matter in living things was indeed different from other matter.

As recently as the second half of the twentieth century Walter M. Elsasser (1904-1991) argued that the existance of living organisms could not be reduced to the forces seen in the rest of nature. He did not accept vitalism at face value though the idea had long been around. Vitalism is the idea that living organisms are caused and sustained by a vital force distinct from the normal laws of chemistry and physics. Elsasser argued that the structural complexity of even a single living cell was beyond the power of any imaginable system to compute and arrived at the conclusion that living and non-living matter are separated by a no-man's land of irrationality. This idea still has some elements of vitalism in it. However, in his book *Atom and Organization*, 1965, he accepts the principle that organisms represent a separate form of matter. Gilson says it this way.

> The mechanistic reformation worked by Descartes first of all demanded the elimination of the philosophical notion of substantial form. . . . Since there are no more forms in the universe of quantum physics, a specific, even general, difference between two immense classes of beings can be explained only by a difference of matters. Our biologist sees this clearly, and this it is that leads him to a "more profound, philosophical" view. "No matter how closely it fits the facts of direct observation", any "theory of organism" will not "be fully satisfactory at a more fundamental level unless it embodies the valid expression of an idea so often emphasized throughout the history of biology, namely, that *organisms* represent a separate form of matter." [2]

Elsasser, as one would expect, had his detractors. He is mentioned here as an illustration that when the best ideas of how to describe the

---

[2] Ibid., Gilson p. 137. Quotes are from Elsasser, *Atom and Organization*.

world—one of which is substantial form or substance—are out of hand rejected one must grasp about for other explanations many of which do not make particularly good sense. One has only to ask how normal elements in food, one type of matter, manage to be changed from that kind of matter to the other kind of matter when absorbed into a living thing? And, when that is done what is it about those normal elements that has changed? Elsasser would have to say the change had something to do with complexity which is not a satisfying explanation.

Evolution vs. Natural Selection

Even though the meaning of the word evolution is hard to pin down, most people today would say it means changes to species that are caused by stresses which are the result of changes in the environment, and that those changes to the species take place gradually over a long time. That is to say, a gray squirrel finding itself in a forest of all black trees would be stressed so much out of fear of being easily seen and, hence, prey for hawks that it would produce black offspring. This is assuming that black squirrels were not already part of the species. However, if one were to mention that black squirrels, thought rare, were already present in the species, they would say that was evolution, too.

Darwin did not use the word evolution until in his sixth edition ten years after this first. There are reasons for that. The first is that in the mid-nineteenth century two somewhat similar words were in use, though they had nearly opposite meanings. One was *epigenesis* which was the doctrine by which an organism grows from a primitive form by the successive acquisition and formation of new parts as the result of survival pressures. As mentioned above, this is the position generally accepted today to mean evolution. The other, *evolution*, was where future species are pre-formed in the seed that need only to develop. Sometimes it is referred to seeds within seeds. Nothing was less like Darwin's thinking than that future species were already present in their ancestors and needed only to make their appearance as the need arose.

To confuse things even more there was the definition of Spencer the philosopher. Recall that he said that evolution was to be regarded as an integration of matter and dissipation of motion that usually was accompanied by other transformations of matter and motion. He was starting with the universe as a whole and working down to galaxies, solar systems, planets, and finally life. Spencer was a well regarded intellectual of the time who fought the whole business of mixing his ideas with Darwin's. Darwin, for his part, had little interest and no need for esoteric ideas like Spencer's. And likewise did not like *his* science which was based on a huge body of facts being confused by the unscientific musings of a philosopher backed up with few if any hard facts.

Darwin's main thesis was that species have changed over time and that these changes came about by the general process he called natural selection. The "transmutation of species" was also a term used at the time and was more in line with his thinking. However, at the time of the publication of his sixth edition in 1869 the word "evolution" was beginning to mean any and all ideas that meant species came into existance without being created. By the time Darwin died in 1882 Darwinism and evolution were virtually synonymous.

Let us return once again to the title to Darwin's book, *On The Origin of Species by Means of Natural Selection; or, the Preservation of Favored Races in the Struggle for Life*. To begin with if "Origin" were replaced by "Continuation" it would be more in keeping with his ideas. The second half of the title suggests that Darwin saw how strongly species resisted change. *Species* is not the correct word as we have seen but to keep the chain of thought intact we will use it as did Darwin. More to the point, all living creatures "struggle for life" whether or not their environment is changing. They must find food every day, survive the extremes of weather, avoid predators and maintain reproduction territory.

The Peppered Moth

It is time to develop the two ideas of how species supposedly came to be in the eyes of the evolutionists—natural selection of already present variability versus stress caused changes. The best way to do that is with a recent example.

In England lives an insect called the peppered moth. It is mostly white with a peppering of black specks. It clings to the bark of trees with a light bark so its coloration is a perfect camouflage to prevent it from being eaten by predators, which are mostly birds. When the industrial revolution came along, the tree bark became black owing to the smoke stack soot. It was known then as it is known now that there always had been solid black members of the species. Over the period of several decades the white moths nearly all disappeared being replaced by black moths. More recently when the environment was cleaned up, and the tree bark became white again, the moths became white again, too. This is held up as a prime example of proving Darwin's theory of evolution. Not really.

What the peppered moth proves, if anything, is natural selection which says that all species have innate variability already present in the population *before* the environment changes. If there are minority members of the population that are better suited to a changed environment those will more readily survive and soon become the majority. Nothing has changed in the species. That is not evolution.

Let us propose another scenario. Assume something happened to the water supply, perhaps a chemical being introduced to it as the result of an earthquake or even a man made change such as in mining. In any case, that chemical causes the bark of the trees to turn, say, red. Any color will do as long as it means that now both the white and black peppered moths are clearly visible on the bark resulting in them all being eaten by birds. Natural selection does not work in that case because there are no red moths to be selected to survive causing the peppered moth to become extinct, at lest in the locality of the changed water supply.

Also, in the case of the peppered moth there were all manner of gray months, something like hybrids, that also appeared along the way as the coloration of the bark changed. That is little else than what we learned about genetics in ninth grade general science. Gregor Mendel (1822-1884), an Augustinian friar, is acknowledged to be the first one to do genuine genetic experiments. He cross-bred peas and among the traits he watched were those with white flowers and those with red flowers. Some of the resulting crossbred plants had white flowers, some red and some pink. The same thing with the peppered moths. The only difference between the peas and the moths was that man intentionally rather than un-intentionally did the cross breeding of the peas. In that vein, Darwin, in his book, *On the Origin of Species*, had quite a lot to say about cattle breeding and other genetic enhancement of various domestic species. From there he as well as others made the leap to saying that what man can do in a few years, nature can do in a long time. The only problem with that is no man has ever made a new species, and nature has never been caught in the act of doing it, either. The peppered moth example is simply a case of *intra*-species variation. It's like now and then a black sheep shows up.

If the peppered moth case showed anything, it was the resistance the moths had to evolving. In fact, all the evidence shows that species are highly resistant to evolution, to the point of becoming extinct rather than changing. If the fossil record tells us anything, it says that species become extinct. And, extinction is something that we have witnessed in modern times. So, in that respect, modern scientific observation does nicely correspond with what the fossil record seems to say. We see that species become extinct, not that they are born.

Another thing that makes the peppered moth case so outrageous is that it happened in a matter of decades. Ask any evolutionist if a new species has ever been seen to from and he will reply in the negative because evolution takes enormously long times. If that is the case how can the peppered moth case in any way be an example of evolution? Yet, it was held up as case where Darwinism was proven to be correct.

Influenza Viruses

At a far end of the spectrum of life, as it were, are the viruses. They have a short life cycle and mutate constantly. For this reason it is almost impossible to come up with a vaccine for the common cold. Yet it is still with us and produces basically the same symptoms decade after decade. On the more dangerous side is influenza. It also mutates rapidly and wildly and can become extremely deadly. Each summer the Center for Disease Control and the Food and Drug Administration take a stab at what the strain of flu will be for the coming winter and has massive amounts of vaccine prepared. Why do we have these diseases? Why have they not mutated into something entirely different after all the millions of years? The reason is that viruses strongly resist mutating into something that they are not. If viruses refuse to evolve when it is accepted that they mutate rapidly and to extremes, why should more advanced species be inclined to evolve? There is even an accepted rule about the mutation of viruses that says they do not evolve. This is shown in the following quote.

> But, the 1918 virus, like all influenza viruses, like all viruses that form mutant swarms, mutated rapidly. There is a mathematical concept called "reversion to the mean;" this states simply that an extreme event is likely to be followed by a less extreme event. This is not a law, only a probability. The 1918 virus stood at an extreme; any mutations were more likely to make it less lethal than more lethal. In general, that is what happened. . . the virus mutated toward its mean, toward, the behavior of most influenza viruses. As time went on, it became less lethal. [3]

---

[3] John M. Barry, *The Great Influenza*, New York, Penguin Group (USA) Inc., 2004, p. 371.

As stated, reversion to the mean is not a law, but a probability. That means it is not confined to viruses or even bacteria. It is what would be expected to happen to all species. The probabilities are that a good mutation will be followed by a bad one to cancel it out. In the above quoted book Mr. Barry describes in detail how wildly successful the flu virus of 1918 became. In that case it was like flipping a coin ten times and having it come up heads every time. Why would it revert to its more normal, far less potent, version other than, using the coin flipping example, the normal probabilities reasserted themselves and tails started appearing? Is not the purpose of evolution for species to be successful? Of course, if the 1918 influenza had continued on its course of becoming more lethal with each generation, it would have killed all possible hosts and itself become extinct. But, not being a rational organism, it would have had no way of knowing that. The reader should observe that even this one example given in the forgoing few paragraphs for all intents cuts the life out of the idea of evolution.

Substance vs. Aggregate

Since we mentioned Spencer and his use of "aggregate" above this is time to elaborate on the difference between a substance and an aggregate. If you mix flour, sugar, shortening, water, an egg and a few other ingredients and bake it you get a cake. In the cake the original ingredients are no longer identifiable so the cake is a new substance. If you mix sugar, oatmeal, other grains, raisins, chocolate chips and maybe a little corn syrup together you can make a granola bar. Here all the original ingredients are still identifiable so this is an aggregate. Or to say it another way, an aggregate is a whole where its describing properties are simply the properties of the individual ingredients. And, the presence of each ingredient in the aggregate does not affect the ingredient. Another example of an aggregate is granite. It is an igneous stone composed of quartz, various feldspars, mica and other minor ingredients that have been forced together by great heat and pressure, but not so much heat as to cause the constituent parts to melt. Each of the separate ingredients is clearly visible especially in polished granite.

It is not possible to give any but the briefest idea of what is at play here. Suffice it to say that starting with the Greek philosophers the concept of substance came into the realm of human thought. Substance is that which a thing is and the accidents of the thing are what we can sense. That is, the orange color of the fruit called an orange is orange. It is an accident of the substance orangness which is the fruit. The color orange does not exist by itself but only as an accident of a substance. It could also be an accident of an orange flower.

Why would Spencer define a species or any living thing, as an aggregate? The answer takes us a step in the direction of discovering why natural selection, evolution, or whatever name one wants to use, was so important to them in the past centuries and still is today. To answer the question one must go back to the sixteenth century and the Protestant Reformation. In Luther's revolt from the Catholic Church he had to dispose of the Catholic doctrine of transubstantiation in the Catholic Mass which depends on substance.

Transubstantiation is that doctrine of the Catholic faith that states that in the little unleavened wheat based host used in the Mass the substance of the host is changed from bread to the body of Christ while the accidents remain the same. That is, there is no possibility of scientifically determining anything different between the pre-consecrated host and the consecrated one because all science can deal with are the accidents of substances and in this case they remain unchanged. Doing away with the concept of substance did away with the central part of what it means to be a Catholic. That is another bedrock necessity of evolution.

This thinking of living matter as an aggregate continues to the present day as evidenced by an editorial in the April 11, 2009 issue of the small journal called *Science News*. In the editor's column the editor was talking about the new Planck spacecraft about to be launched. It was designed to measure the background radiation in space. [4] Max Planck was the physicist who first formulated quantum mechanics. Here is a paragraph from the editorial.

Back then, a century ago, nobody realized that Planck's quanta

---

[4] The data collected by the Planck spacecraft were not at all what was expected. To the chagrin of the investigators it showed that the earth could be the center of the universe. So unacceptable were these results that another more sensitive spacecraft was launched and it only made the data more precise. Yet, a third probe was launched only to give the same results. The radiation is unequally arrayed about an axis that passes through the earth. So hateful was this finding that scientists have dubbed that axis "the axis of evil." It is worthy of note that it is a symptom of how deeply ingrained is the idea of evolution that the scientists had no problem getting congress to appropriate money for the second and third tries.

were the key to understanding the universe itself. But, now cosmologists "know" (physics jargon for "strongly suspect") that quanta from the beginning of time generated seeds of matter that grew into the galaxies that populate the cosmos today. Explaining the origin of the galaxies, the stars, and star products (such as people) and even, perhaps, the universe all depends on mapping the quantum signatures written in the radiation that the Planck spacecraft will scrutinize. [5]

At least he was willing to say that physicists are not all-knowing in their physics and would stoop to using the term "strongly suspect" rather than "know." What he did for physicists he would not do for evolutionists, though. Next he says humans are little more than star flatulent. But, do not miss the point. If interstellar gas evolved into bacteria, to monkeys, to us, it means categorically that humans do not have spiritual, immortal souls. That means there is no final accounting, no judgement, no life hereafter. That, in turn, means we can be as murderous, as adulterous, and generally licentious, as we please. Forget about everybody else. The editor of *Science News* could say without fear of contradiction that we came from a cloud of interstellar gas. We as humans are the same as a cloud of gas it is just that we look a little different. That is, to him we are nothing but granola bars.

Darwin, Evolution and the Assassination of God

So why did Darwin become so totally associated with evolution? There were reasons other than saying natural selection, that is, Darwinism, and evolution deal with the same general subject. At the time he published his first edition in 1859 there were those who agreed with Darwin and those who agreed with Lamarck and Spencer and most people who agreed with none of them. But in the end, especially by Darwin's later life, they all had one common objective which was to do away with creation. That is, everybody made common cause in getting rid of God.

The book written by Robert T. Clark and James D. Bales called *Why Scientists Accept Evolution* has already been quoted. It is mentioned here to point out that the authors come to the subject from the point of view of

---

[5] Editor's Column, *Science News*, April 11, 2009.

quoting what the actual people in the nineteenth century said about evolution. An estimated seventy-five percent of the book is direct quotes. Many of the quotes are from letters—that still exist—because people tend to be more candid in letters than in published material.

The first thing one sees is that all of them Hutton, Lyell, Darwin, Spencer, T. H. Huxley, Wallace and others first lost their faith in the Christianity of their youth which for them was the Bible since being Protestants they followed the dictum of *sola scriptura*. But, even that is not strictly true. They first came up against the problem of six days of creation as given in Genesis. The six days were taken literally at the time. They knew about fossils and it seemed to be unscientific to say the universe was only a few thousand years old. This led to a gradual and general loss in faith in the supernatural which in turn led to rejecting most if not all of the Bible.

This is significant because it is not the case that they were believers in creation and were forced to abandon creation and the Bible in favor of the great new idea of evolution as an intellectual necessity. It was the other way around. To a man, they lost their faith in creation first and then were looking for something to replace it—anything. Along came evolution. In one way or another they all admitted it was a dumb idea but it was all they had. It is astounding how many times they use the word "believe" in their writings. They all knew and said that the hypotheses of evolution was not nearly far enough along to be called a theory. They also all expressed the *belief* that future generations would find something that would put the hypothesis on a firmer footing. This has not happened.

A further word about creation. The personalities listed above under the heading of "Fathers of Evolution" generally accepted that it was unreasonable to assume the universe always existed. Some of them accepted creation on some level. All that was necessary to make the original universe or a new species was something outside of nature. In the present day if scientists succeed in making a new species in the laboratory, as they are now intent on doing, it could very well be a new species since human intelligence is part of the human soul, a spirit, and is outside of nature. They are trying to make a new spices to prove evolution but all they will do is prove creation or in current parlance, intelligent design.

A word about intelligent design. In many circles that idea is not helpful in debating evolution. That is because the earth is commonly referred to as "mother earth" by environmentalists who naturally are also evolutionists. "Mother earth" could be nothing but another term for pantheism.

These "fathers of evolution" were really agnostics, most of them readily admitted it. They could not get off the horns of the dilemma: creation in six days or life and species simply popping up out of dust and rocks over a long time. They write many times about how they were torn, in fact, muddled. As an example we find that Darwin wrote a letter to Asa Gray [6], November 26, 1860 saying he could not bring himself to accept design as Gray did.

> But I grieve to say that I cannot honestly go as far as you do about Design. I am conscious that I am in an utterly hopeless muddle. I cannot think that the world, as we see it, is the result of chance; and yet I cannot look at each separate thing as the result of Design. To take a crucial example, you led me to infer (page 414) that you believe "that variation has been led along certain beneficial lines." I cannot believe this; and I think you would have to believe, that the tail of the Fantail was led to vary in the number and direction of its feathers in order to gratify the caprice of a few men. Yet if the Fantail had been a wild bird, and had used its abnormal tail for some special end, as to sail before the wind, unlike other birds, every one would have said, "What a beautiful and designed adaptation." Again, I say I am, and shall ever remain, in a hopeless muddle. [7]

A year later Darwin again expresses his uncertainty to Asa Gray, September 17, 1861.

> I must think that it is illogical to suppose that the variations, which natural selection preserves for the good of any being have been designed. But I know that I am in the same sort of muddle (as I have said before) as all the world seems to be in with respect to free will, yet with everything supposed to have been foreseen or pre-ordained. [8]

Here is where we are. To accept evolution you must believe in uniformitarianism which is at odds with the geological record. However,

---

[6] Asa Gray, famous for his work on botany of Northern United States.

[7] Charles Darwin, *The Life and Letters of Charles Darwin*, Edited by his son Francis Darwin, Vol. II, p. 201.

[8] Ibid., p. 232.

uniformitarianism is a point these men who began the modern idea of evolution all demanded. Then there is that odd thing about "short-time" being normal and "long-time" being magical. This in the face of the fact that no one, especially no scientist, has the slightest hint of an idea of why "long-time" has those strange properties. It is not lifeless matter that makes life or new species because we've all been watching that for millennia and nothing has shown up. To believe that you must also believe that you can get something for nothing, that is, the greater comes from the lesser. If you've figured that out, your bank account must be enormous. These are some of the doctrines that must be believed in order to get away from creation.

In the novel *Hubot-Human Robot* by Joseph P. Cody (yes one and the same), 2012, there appears a somewhat whimsical conversation between a young man, Ed, and a well educated twelve year old boy, Jason, on page 282. They have recently met and are whiling away time as they are inching along in traffic due to a snow storm. The subject of evolution comes up. Ed says,

". . . Have you ever planted a garden?"
"I helped mom plant the flower beds a couple of times."
"Did flowers grow?"
"Of course."
"Would you ever think of planting small pebbles and expect flowers to grow?"
"That's silly. Of course not."
"That's right. It is silly. But, that's exactly what you are saying when you believe in evolution."
"Nah. Evolution takes millions or even billions of years."
"Aha. What you are saying is that something that is manifestly absurd in a short time is not only possible, but entirely expected in a very long time. This means that you believe in God and you believe in creation, only, what I call God, you call time."

If you mentioned something like that to a modern scientist he would hit you. Please understand that *Hubot-Human Robot* is a novel and is not a book about evolution.

Chapter 4

Evolution and Moral Society

What was going on in Europe and specifically England during the last of the eighteenth and all of the nineteenth centuries to cause people's thinking to become so fuzzy? What follows will be a discussion of the society out of which the hypothesis, now generously called a theory, of evolution sprang. In a previous section the point was made that all of the "fathers" of the modern evolutionary movement had lost their faith in the Bible and hence creation and were in search of something to replace it. But, what was the cause of this loss in faith in Christianity and the Bible?

Before we start that story, a few words should be said on the subject of loss of faith in the Bible. Many significant people who bought into the notion of evolution were Bible believers who switched to evolution because it was more compelling to them, or so many of their biographers, incorrectly, say.

For example, Karl Marx (1818-1883), was born of Jewish parents but was baptized a Lutheran in 1824 because his father had to convert to Lutheranism to maintain his profession as a lawyer. Karl attended Lutheran schools and was deemed moderately proficient in theology. But he became interested in revolutionary movements and became an atheist and a materialist, rejecting both the Christian and Jewish religions. He got a doctoral degree in Berlin in 1841. Within a year after the publication of Darwin's first edition in 1849 Marx read it. Marx was well aware that to develop a scientific outlook on society—which was the only way that the emerging movement of the working class could establish socialism—a historical approach was needed. In 1861 he wrote: "Darwin's work is most important and suits my purpose in that it provides a basis in natural science for the historical class struggle."[1] Notice that Marx had long been an atheist when he read *The Origin of Species*. As a result, it was the general apostasy of the times as well as his personal apostasy rather than evolution that caused him to lose interest in the Bible and religion.

Another case was Joseph Stalin (1878-1953). As a theology student studying to be a priest in the Russian Orthodox Church he became interested in the nationalist movement in his native province, read Victor Hugo's writings on the French Revolution as well as Karl Marx' writings. And, by the way, he also read Darwin. "Despite his Christian upbringing, he had become an atheist after contemplating the problem of evil and learning about evolution through Charles Darwin's *On the Origin of Species*."[2] In Stalin's case he saw the evil and poverty around him and was deeply into a revolutionary mindset before he encountered Darwin. The same is true of others.

The Middle Ages

Now to our story. It starts with the Middle Ages. In the early part of that period the rational mind began to assert itself as the Greek philosophers were rediscovered. This can even be to some extent laid at the feet of the Scholastics (a term which means Schoolmen). St. Thomas Aquinas (1225-1274) did a lot to show how the doctrines of the Catholic faith were in harmony with reason. Reason was not to supplant faith, but help make revealed truths more vibrant in the minds of believers. There was nothing wrong with that.

This led to the Renascence, 14th, 15th, and 16th centuries, where people like Copernicus (1473-1543) proposed heliocentricity where the earth revolved around the sun and the earth revolved on its axis. Then, Galileo (1564-1642) demonstrated heliocentricity with his telescope by observing the orbits of the moons of Jupiter. Things like this were changing the way people thought about the world around them.

> "Experiment" says Leonardo da Vinci (1452-1519) . . . "is the true interpreter between nature and man." Experience is never at fault. What is at fault is man's laziness and ignorance. . . . This was

---

[1] http://www.marxists.org/archive/marx/works/1861/letters/61_01_16.htm
[2] Simon Montefiore, "Epilogue". *Young Stalin*, 2007, Britain, Weidenfeld & Nicolson. p. 395.

the essential note of the new European movement; it was applied science, not abstract, speculative knowledge, as with the Greeks. "Mechanics," says Leonardo again, "are the Paradise of the mathematical sciences, for in them the fruits of the latter are reaped." [3]

Early in the 16th century the Reformation made its appearance. The Renaissance and the Reformation are not the same. ". . . that revolt against medieval culture which is the Renaissance is by no means to be identified with the revolt against medieval Catholicism which is the Reformation." [4] However, "It is no accident that the Reformation took place in Germany where Church and State were the most inextricably entangled." [5]

Martin Luther (1483-1546) ignited the Reformation which produced Protestantism by nailing his 95 thesis on the church door at Wittenberg in 1517 in an act that would change Europe forever. This movement changed people's thinking on their life in this world, and the world to come back to the way it had been before the Christen era, in fact, back to before the Greek philosophers. John Calvin preached that certain men—his closest followers to be exact—were chosen by God as the elect destined to go to heaven at the expense of the huddled masses. Speaking of Calvinism Hilaire Belloc makes the following observations.

> The novel object was an implacable God, the appetite was love of money. There is a dark instinct of horror which is found lurking or patent in all antiquity and modern pagan ritual, a demand for victims and a prostration before dreadful power. Calvin provided victims. For his ardent disciples, remember, were elect. It was the others who were damned, and as for love of money, a philosophy which deride good works and dreaded abnegation [was] let loose in all its violence. Calvinism had men enrich themselves, and they have done so. [6]

---

[3] Christopher Dawson, *Dynamics Of World History*, La Salle, Illinois, Sherwood Sugden & Company, 1978, p. 123.

[4] Christopher Dawson, *The Formation Of Christendom*, New York, Sheed & Ward, 1965, p. 290.

[5] Ibid., p 288.

[6] Hilaire Belloc, *How The Reformation Happened*, Rockford, Illinois, Tan Books, 1928, p. 80.

With Luther's rejection of reason and demanding salvation by faith alone, and Calvin's bent toward worldly wealth and gain, Protestantism took the transcendent out of the message of Christ. Man lost his destiny to reach beyond this earth and beyond this life by means of his cooperation with the grace of God. Rather, Protestantism became a model of how to live as comfortably as possible here on this earth in this life, the only life.

Then in the 17th and 18th centuries along came the Enlightenment and Newton (1642-1727) where everything was put on a scientific basis. It is interesting to note that making something scientific is to convert it to numbers and equations. This process of rationalizing life tended to make faith even blander than it had already become. The thriving, palpable faith that was Christianity of the 11th, 12th and 13th centuries began disappearing and never again produced the stable, sane society of that age.

Here one must recognize a strange dichotomy that appeared in the 18th century. At the time, science, à la Newton, was putting reality on a strict scientific basis where everything could be reduced to formulas, numbers and calculations. That was all very much a structure of reason. On the other side there appeared Romanticism which was a sweeping reaction against reason, science, order, authority, tradition and discipline that convulsed Western Civilization in the late 18th to the mid 19th century. In art this was overlaid by that is called the Romantic Period which lasted from 1800 to 1910. This period was characterized by emphasis on the expression of individualized and subjective emotion coming from the sensitivity of the mid 18th century music, art and literature as well as by a sense of the exotic, mythical and supernatural.

The Nineteenth Century

This brings us to the mid-nineteenth century. The natural curiosity of man could not let go of the thing called experimental science that was made all the more possible by the great strides in mathematics. However, the legacy of the Reformation was upon Europe. Protestantism was as much as anything a reaction against the world of the scholastic philosophers were reason controlled will, and made a world where will controlled reason.

In Germany, Richard Wagner (1813-1883) was intent on revolution— will over reason—on destroying the Christian society of Europe and especially in Germany where he lived and worked. He saw the Classical/Christian tradition with is strictures on moral behavior as killing the freedom of the human person. His first attempt was in backing a political uprising in Germany in 1848-1849 which failed. His association with the revolt cost him his job as director of Dresden's orchestra. Seeing that political revolution as too unpredictable, he set about his mission through culture, particularly that of music.

> Monarchy was the rule of one man in the political sphere; melody in music was the domination of a row of notes by a tonal center. By the mid-nineteenth century, monarchy had come to be seen as *the* form of government sanctioned by the Christian culture. Melody, as the coherent organization of notes around a tonal center that dominated and organized its emotion, had become the prime expression of music. Melody was the soul of music; it was to music what plot was to tragedy, according to Aristotle. . . . Wagner had become radically dissatisfied with both forms of order. [7]

Wagner wrote his first opera *Die Feen* in 1833, but it was his second one *Das Liebesverbot* that was his first to be performed, and that in 1836. In his early career his revolutionary bent of mind was not evident in his music though his sexual life was always dominated by illicit affairs. In his opera *Tannhäuser*, written between 1843 and 1845 and first performed in the fall of 1845 he presented a tearing between lust and virtue. Tannhäuser leaves a safe but servile erotic life with the goddess Venus. Then he joins the real world where all his attempts at virtue fail. At the end he is resolved to return to Venus but dies before he can.

Then in his famous opera *Tristan und Isolde*, (written in the period of 1857-1859 and first performed as an opera in 1865) Wagner set a new standard in sexual immorality. Part of the impact of this opera was Wagner's use of the chromic scale, the common scale on any piano today. That is, it has twelve half notes per octave including all of the white and black keys. But, with it he produced a new tonality, one of adventurous and troubled dissonance, that appealed to the lower appetites in a fundamental and powerful way. Since opera is a sung play, the lyrical text was integral to providing a bearing and objective to the total effect. It was saturated with sexual innuendo and the outright lust of passionate adultery. The music continues for hours in a feverishly unanswered tonal ten-

---

[7] E. Michael Jones, *Dionysos Rising*, South Bend, Indiana, Fidelity Press, 2012, p. 31.

sion until the very end.

The doctrine of Christian love, initially so promising, is freed from its narrow confinement inside the Christian moral law and can now expand to embrace all mankind in an orgiastic, enravishing announcement of sensual emotions. Without the text, which is to say, without the will of the poet, which imbue the music with both direction and meaning, the absolute musician . . . swam aimlessly and restlessly to and fro . . . . [8]

Wagner's total break from tradition in *Tristan* appeared at the exact time when it would be most easily received and cause the most damage.

The West would meditate with increasing intensity over the next century and a half on the price it was willing to pay for sexual liberation, liberation from the light of reason, liberation from the cultural achievements that were based on society's subjugation of individual passion to the order of reason. Was it worth it after all? The audiences that found themselves awash in the tepid chromatic modulation of *Tristan* could ask themselves this question over and over again. [9]

Fredric Nietzsche (1844-1900) at the age of seventeen heard the music of *Tristan und Isolde* in the score for piano and went a little off the deep end. At the age of twenty, Nietzsche went so far as to go to a brothel and deliberately contract syphilis as an act of rebellion against the existing society and in so doing made a pact with the devil. He suffered with his disease all his life finally succumbing to dementia in his last years.

Luther hated reason every bit as much as Nietzsche and, in trying to make up the worst insult imaginable, [Luther] came up with the formula "Reason is a whore." Nietzsche raised by his Protestant pastor father in the sanitized Lutheranism of the nineteenth century, took the Lutheran anti-rationalism to unheard of peaks of fanatical piety. He absolutized the hate for reason; he provided a cultural program for banning the entire tradition of the Chris-

---

[8] Ibid., p. 36.
[9] Ibid., p. 41.

tian/Classical West, which was based on a reverence for reason, whose cause he traced back to Socrates. In *The Birth Of Tragedy*, Nietzsche's first book and the one he wrote while under the maximum influence of Wagner, Nietzsche proposed a "new dichotomy,": "the Dionysian and Socratic principles." The advent of reason among the Greeks signified the death of music and tragedy, the banning of worship of Dionysos. . . . It is also the source of the tradition of world harmony in the West. . . . With the advent of Socrates, music was subjected to reason. In order to end the Christian/Socratic age, reason would now be subject to music, which is, according to Schopenhauer, the immediate expression of the will. [10]

It is true that Martin Luther rejected reason, but he also denied freewill. For him man was saved by faith alone. Nietzsche and Luther had common ground in their dismissal of reason, but Nietzsche far from denying freewill made will supreme. The book *The Will to Power* was compiled from Nietzsche's notes written between 1883 and 1888, and first published in 1901 a year after his death. In it he argues that a person's actions toward other people are only to exercise his will over those people. The same goes for one society toward another or one country toward its neighbors. This left willful man free to engage in indulgences of all kinds, especially sexual.

Wagner the revolutionary *manqué*, had come upon a musical invention in *Tristan* so powerful that it would provide the vehicle for overturning the hegemony of Christian culture in the West. The twenty-four year old Nietzsche [this is 1868 the year of the last edition of *The Origin Of Species*] would write the operator's manual [*The Birth Of Tragedy*] for this soon-to-be-born cultural revolution that would herald an age of sexual excess and unfettered will. That manual would bring to the West the ceremonies of lust and cruelty that had been lost since the ascendancy of Socrates. [11]

Nietzsche was bent on hedonism at the expense of all else.

Just as under the old dispensation, salvation depended on subordinating the individual will to the will of God . . . the new civiliza-

---

[10] Ibid., p. 55.
[11] Ibid., p. 56.

tion was to be an inversion of that ideal. Now everything was a function of will, and salvation was achieved by pursuing the will as single-mindedly as possible. [12]

This led inexorably to where the abnegation of reason always leads, namely to the demonic.

The whole notion of the will pursued to the point of violating nature as a way of becoming privy to her secrets is a profoundly demonic idea. And it was that idea that lay at the heart of the new civilization Nietzsche was proposing in *The Birth of Tragedy*. [13]

Needless to say, the writings of Nietzsche have had a profound affect on the modern world and would be called by the name of modernity. An excellent treatment of these people in this period is given from the quoted *Dionysos Rising*.

Tristan und Isolde was performed in Europe and later in America. The average man in the street would not have heard it, but the movers and shakers in the evolution movement would have. That particular opera was, of course, not in itself the only reason for the acceptance of evolution nor were the writings of Nietzsche or the libertine sexuality that it helped promote the only cause. Other cultural factors were involved. It will not be gone into here other than to mention it, but capitalism profited immensely from evolution. With it the rich could excuse their greed by saying they could not help subjugating the poor in their sweatshops and dirty factories. It was a case of the survival of the fittest, and they were simply more fit.

All of this is presented as a way of saying that the hypothesis of evolution did not spring out of thin air or from the detached reasoning of scientists, it was a product of its day. Darwin was quietly going about his business of studying bugs, barnacles, and birds. He was a scientist trying to make sense out of the physical world as any scientist should be doing. Meanwhile, Wagner and Nietzsche as well as a host of others were creating chaos in society so a licentious, lustful sexuality could come into being. The latter succeeded in causing the Church's teaching on morality to be cast off most specifically among the educated classes. This meant the Biblical account of creation was gone and a substitute was needed

---

[12] Ibid., p. 58.
[13] Ibid., p. 60.

and that was evolution. It swept Darwin's work up into it, much to his dismay, to give it a fig leaf of respectability in the scientific community and society at large. With the passage of a few generations, evolution, now incorporating Darwinism, was uncritically taken for granted, and it, in its turn, was used as the "scientific" basis for rejecting Christian morality. There is nothing scientific about evolution and there never was.

As one would expect, evolution was not accepted by everyone at one time. It took the entire last half of the 19th century for that to happen. But, that time period was pivotal in letting it become entrenched. The music of that period, which has evolved (societal evolution in this case) into the acid rock of the Rolling Stones and like groups, and the theory of evolution are now connected in a pact of death with the devil. At least Satanists are more honest that the rest of us. They accept that there cannot be unbridled sexual license without human sacrifice. Why is everyone so surprised?

Society at large sees sacrificial death today primarily in abortion, but increasingly common as euthanasia. This says nothing about suicide and all kinds of self destructive behavior short of that. This includes drug abuse and irresponsible sex that leads to AIDS. Why do school shootings seem so terrible? A minimum of nine out of ten are the result of students who had been placed on mind altering drugs such as Ritalin. Added to this are hours a day spent playing video games where the object is killing people. Murder is murder whether of students and teachers or millions of unborn babies. Terrorism is more fanatically carried out in the sex education programs in our public schools than by suicide bombers. As shown above and as will be developed more further on, the root cause of the modernist mess we find ourselves in today is the result of letting will control reason.

If things were coming apart in Europe in the nineteenth century, it was at least as bad in America. A little know feature of the American Experience is the way sexual license was set free here in that century.

Imagine an America where pornography was not only a flourishing trade, but ubiquitous, even handed around in the public schools at the time; where contraceptives were not only sold, but in high demand; where books advocating birth control, and describing the most popular contraceptives, sold in the hundreds of thousands; where abortionists—and abortion—were common, perhaps even more common than today (the two-child family, the norm for

Protestants, was often maintained by recourse to abortion); where nearly one in four adults was afflicted with some sort of venereal disease, due to prostitution and public tolerance of brothels.

Imagine that, in the same period, in New York, one in four married couples was childless; in three-fourths of married household, there was only one child; cohabitation was common, and, perhaps, as many as 20% of adults never married; divorce was acceptable, and routine; where public schools were closing for lack of children.

Can you give a guess when this was? It was during the decades before and after the Civil War.

Those are some of the astonishing facts marshaled by Dr. Allan Carlson in his latest book, *Goodly Seed: American Evangelicals Confront Birth control, 1873-1973.* [14]

Largely through the work of Anthony Comstock national and state laws were passed outlawing pornography, contraceptives, and abortion. Comstock's political descendents would be the socially conservative Religious Right of today.

It would not be surprising if this sort of moral behavior was also the norm in Europe at the time. One decade before the Civil War would be the year 1850. Darwin published the first edition of his book *On The Origins of Species* in 1859 just one year before the Civil War in America began.

---

[14] "From The Mail", *The Wanderer*, St. Paul, Minnesota, January 31, 2013.

Chapter 5

Physics, Economics and Evolution

A compelling case can be made that connects physics, economics and evolution with the English ideology that developed in the seventeenth century. The common thread of these things is capitalism which is a system of strife and struggle. In capitalism, everyone is competing with everyone else producing winners and losers. The political system we, in America, inherited from the English is one of strife—winners and losers.

Isaac Newton (1642-1727)

Newton lived and worked in the last half of the seventeenth century and early eighteenth. He introduced the idea of strife and struggle into physics. Before Newton, in the Scholastic philosophy, force and motion were driven to an end or goal called *telos* since all was driven by Divine Providence. But, the Protestantism of the period sought to do away with God. Never mind what they say, God was not high on their list of priorities. Newton had the answer in that any body continues in the state of rest or uniform motion (a state resulting from its inertia) unless acted upon by a force. That is, things moved or did not move based on competing forces. A rock sat on the ground being held in place by gravity. If you lifted hard enough on it your competing force won and gravity lost.

It should be noted that Newton was appalled by the thought that gravity would act by an unseen and unexplained agent. He felt that in time we would learn that there were particles or something traveling between two masses. When he wrote his work, *Principia Mathematica,* (1687), in which he presented his principles of motion and gravity, he used obtuse formulas and notations. This was mainly because "In formulating his physical theories, Newton developed and used mathematical methods

54

now included in the field of calculus, expressing them in the form of geometric propositions about 'vanishingly small shapes.'" [1] That is, he, among others, were inventing calculus, probably the biggest advancement in mathematics since the zero was introduced into mathematics in Europe in the thirteenth century. As such, there were no standard symbols and conventions yet established for this new form of mathematics. Nonetheless, that meant that few people of his time would understand what he was proposing and he liked it that way so no one would question him on such an outrageous idea.

Newton expresses his unease with the idea of two bodies attracting one another with nothing passing between them in a letter of February 1693. Here Newton is answering a letter from Richard Bentley. Following is part of Newton's letter.

The last clause of your second Position I like very well. "Tis unconceivable that inanimate brute matter should (without the mediation of something else which is not material) operate upon and affect other matter without mutual contact; as it must be graviton in the sense of Epicures be essential and inherent in it." And this is one reason why I desired you would not ascribe innate gravity to me. That gravity should be innate inherent and essential to matter so that one body may act upon another at a distance through a vacuum without the mediation of anything else by and through which their action or force may be conveyed from one to another is to me so great an absurdity that I believe no man who has in philosophical matters any competent facility of thinking can ever fall into it. Gravity must be caused by an agent acting constantly according to certain laws, but whether this agent be material or immaterial is a question I have left to the consideration of my readers. [2]

It is worth noting, at least in the opinion of this author, that in the above extract from Newton's letter he does not accurately quote Bentley's letter. The exact quote is as follows: ". . . tis unconceivable, that inanimate brute matter should (without a divine impression) operate

---

[1] https://en.wikipedia.org/wiki/Philosophi%C3%A6 Naturalis Principia Mathematica#cite note-gschol-hnf-10

[2] Original letter from Isaac Newton to Richard Bentley dated shortly after 2/18/1693, 189.R.4.47, pp. 7-8, Trinity College Library, Cambridge, UK. Published online: October 2007.

upon and affect other matter without mutual contact: as it must, if gravitation be essential and inherent in it." [3] Notice that Newton cannot bring himself to repeat "divine impression" and instead says "something else which is not material." Newton, as with many others that have formed our modern thought, was an atheist.

However, even today we do not know how gravity works. We have Newton's equation of the inverse square law, but that only describes *what* happens, not *why* it happens.

There is a growing acceptance of what is called zero point energy. Essentially that means that even at absolute zero atoms still vibrate which means there is still energy about. Some theories say this remaining energy is nothing of consequence and some say it is immense. The later position is bolstered by the also growing acceptance of dark energy in the universe. This is postulated because there does not seem to be enough visible matter and energy in the universe to account for what astronomers see. This dark energy hearkens back to the pre-Einstein world (see the section on Modern Physics below) where ether was thought to be the "substance" of empty space. If dark energy—ether—is admitted and that it is actually denser than the atomic particles that make up matter—protons and electrons—there exists a reasonable explanation for gravity.

Since atomic particles are less dense than ether, yet occupy a definite position within the ether inside the atom, this means that the total density of the ether within the atom will be less than the density of ether outside the atom. This imbalance will cause an ether vacuum between the inside and outside of the atom. Since nature abhors a vacuum, the ether will seek to distribute itself in order to eliminate the vacuum. *In short, ether's effort to eliminate the vacuum is the cause of gravity.* That is, the less-dense ether inside the atom will attempt to draw in the denser ether outside the atom. This vacuum force will continue until equilibrium is reached, but, in fact, equilibrium is never reached, and thus the force of gravity between the two objects persists indefinitely. Emphasis his. [4]

---

[3] Source: Original letter from Richard Bentley to Isaac Newton dated 2/18/1693, 189.R.4.47, pp. 3-4, Trinity College Library, Cambridge, UK. Published online: October 2007.

[4] Robert A. Sungenis and Robert J. Bennett, *Galileo was Wrong*, Vol. II, 10th edition, State Line, PA., Catholic Apologetics International Publishing, Inc, 2014, p. 25-26.

Adam Smith (1723-1790)

There is nothing wrong with the idea of capitalism. It just happens that every place we've ever seen it applied it has been unbridled capitalism. That has been condemned by most of the Popes starting with Leo XIII. That said, Karl Marx and Darwin knew of each other. For a time they lived twenty miles apart in England, though it is not known if they ever met. Adam Smith published *Wealth of Nations* in 1776 and died in 1790 so clearly Darwin did not know him. But, this is said about the two: (1) Darwin was a late product of the Scottish Enlightenment; (2) Darwin was influenced by Adam Smith's ideas far more than is apparent from his one mention of Smith; and (3) The theory of evolution is remarkably compatible with free enterprise economics, especially regarding the "invisible hand."

Smith's famous "invisible hand" statement is as follows so we see it clearly. "By directing that industry [the person's work] in such a manner as its produce may be of greatest value, he [the worker] intends only his own gain, and he is in this, as in many other cases, led by an invisible hand to promote an end which was no part of his intention." [5]

Charles Darwin (1809-1882)

As noted above, Darwin said he found the basic theory of natural selection not from nature but from Malthus' book on economics. We see a progression of thought here. From Newton we have the "unseen hand" of forces, especially gravity, that are the basis of physics. Of course, formulas like $F = ma$ were operative before Newton, but it is the way one understands reality that is significant. From Adam Smith appears the "unseen hand" of economics doing "what is best for everyone." From Darwin we see the "unseen hand" of natural selection making new species "that are best." What is all of this reference to the "unseen hand" but Divine Providence?

Isaac Newton, Adam Smith, and Charles Darwin were all products of the enlightenment that had as one of its main goals to get rid of any notion of God and religion. Darwin's theory of evolution gave a pseudo-scientific explanation to economics and sociology that left out God. That is why Darwinism was important then, and is still important today. Science is distinctly a secondary aspect of it.

---

[5] Adam Smith, *Wealth Of Nations*, 1776, Book IV, Chapter II, p. 430.

Chapter 6

The Human Body and Survival

There is the persistent idea, especially in the Catholic Church that it is alright to think that the human body evolved to the point where we see it now and then at some point God created the human souls for the first man and woman thereby giving them a spiritual soul with intelligence, that is, the ability to reason and free will. There is no allowance for the soul to have evolved since being a spirit it was created as a singular act of God for the first humans and for every one who followed. With that in mind it is interesting to see how that pre-human would have fared with respect to survival, to say nothing of evolving.

The Human Body

We start by assuming that evolution happened just as the evolutionists say. They accept that the human body has shown no detectable evolution since the first tool users appeared on the fossil scene. The question remains, how did our present bodies survive without a rational mind? It is reasonable to say that it would have gone extinct in a single generation if not a matter of days or weeks.

Start with a mental exercise where you are hungry and do not have a rational mind that allows you to make a spear or even throw a rock at all let alone with enough accuracy to hit and kill an animal. You are not a tool user; you are what we call a dumb animal. Have you ever tired to catch a rabbit? Don't forget you cannot use a trap, set a snare or shoot it with a bow and arrow, to say nothing of a gun. That means you must catch it either by running it down like a fox might or lying stealthy in wait for it to scamper by and pouncing on it like a cat would. And, if you were able by some chance to lay your hands on it, would you be able to kill it before it bit and scratched you so badly that you were forced to let it go? Even with legs five inches long a rabbit can run faster than you can. You do not have claws or fangs. You can't even climb trees well.

The point is, you would never catch enough rabbits to stay alive. Add to that, people who have been forced to live off the land say that a man will starve to death trying to live on rabbits—even with a gun. Larger game is needed to keep a man alive and they are much harder to capture and kill than rabbits. That is to say a human body with a monkey's brain would not last long.

So, how did the animal that existed the moment before God gave us a rational mind survive? Besides being entirely noncompetitive in the area of catching protein for a normal diet, the human body has no natural covering. Other warm blooded animals have coverings of either feathers or hair. We will leave aside cold blooded creatures with scales as a third known covering. Humans have neither hair nor feathers. Without the intellect with which to make clothing and shelter, there would have been only a few places on earth at or near the equator where non-intelligent humans could have survived. Even at that, all ecosystems are teeming with competing species, of which the human body would be the least competitive when vying for food and "nesting" territory.

The human body can digest a wide variety of foods including many vegetables. That is, humans are both herbivores and carnivores, that is to say omnivores. But living only on fruits, nuts and berries means starving most of the time. In the modern world it is possible to live as a vegetarian, but in the wild with no grocery stores, it would be rare indeed.

It might be objected that in an equatorial jungle there would be plenty of food growing close at hand. It may not be that simple. Even if there were bananas growing wild monkeys are far better climbers and would get them as they ripened before a man had a chance. And, do not forget, there will be no throwing stones to drive the monkeys off.

After that is the problem of man as prey for other predators. We would have our teeth and our claws, that is finger nails, with which to defend ourselves. Not much of a contest with a tiger. There was a *Far Side* cartoon in which two alligators were lying on a beach under a palm tree with obviously full bellies. Scattered about were bits of clothing, a camera and sunglasses. One said to the other, "That was great. No claws, no fangs, just soft and pink." How about that for survival of the fittest?

Endurance Hunting

One of the most common arguments put forth for man's ability to survive without intelligence is his ability to sweat. The human body can sweat more than any other mammal. That along with the lack of hair makes it possible for him the shed heat the best thereby making him able to run longer than any animal, though in a sprint he will generally lose. It is argued that on the savanna, as long as he can keep his prey in sight, a man can run down any animal. The animal having a hair covering cannot dissipate the heat fast enough and will fall down in heat exhaustion so the man can dispatch him with ease. This is called endurance hunting. Exactly where the term originated is uncertain but some evolutionists claim that it along with an opposable thumb and a few other traits make it so intelligence had no significant survival advantage for humans.

That is a perfect example of a case where the idea sounds great in the abstract but it is the details that bring it down. The first is the part about keeping the prey in sight. On any open land whether a savanna of Africa, or the planes of North Dakota, the land is *not* flat. That means that in a half mile the prey will go over a rise or into a declivity and be lost from sight. But, it will be argued, the man could track it. The sense of smell of a human is not well enough developed to track an animal by its scent, and dumb animals do not track by looking for characteristic impressions in the ground. Even if you allow for tracking by sight, tracking while running is difficult. Meanwhile, the animal will have managed to have gotten over the next ridge and have had time to rest and cool down for its next run if the man persists and can locate it again.

One must ask why animals dissipate heat so poorly, that is, why don't they have the capacity to sweat better? It is precisely because sweating wastes so much energy and the conservation of energy has a huge survival advantage. What that means is the man will have expended a large percentage of the energy value he would have gotten from eating the animal had he captured it. Then, what if he expends all that energy and does not kill the beast after all? Will he have enough strength to succeed the next time?

Then we come to the point where the animal has fallen down from heat exhaustion and the man has merely to "dispatch" it. How? A cougar after bring down a deer by digging it's claws into its back will instinctively clamp its jaws on the deer's throat thereby suffocating it and if possible breaking its neck. The thought of a man doing the same thing is laughable. The only recourse he has is to strangle it with his hands. We might allow that he is a "tool using animal" (more on that later) and might hit it in the head with a rock. However, the odds are there would be no rock conveniently at hand. Then, do not forget that all animals have a reserve of energy for a last effort to survive. Anyone who has not been kicked by a hoofed animal does not understand how dangerous even a small animal can be with its hooves, to say nothing of its bite.

Assuming now that the man has killed his deer, he must eat it. No tools are allowed especially no knives or even sharp stones. Gnawing through the hide is a project all in itself. Remember, shoes are made from that material. And, he must plan to take as much of the booty as he can conveniently carry all those miles back to his family. That is to say nothing about his being extremely thirsty after all that running and sweating.

Additionally, what happens when our persistent hunter finally overcomes his deer, it is dusk, the wind picks up and it starts to rain. Now, the ability to shed heat works against him and he is in danger of dying from hypothermia. Having hair, other animals strike a balance between shedding head and retaining it. They even have the ability to make their hair lay flat against their skin to shed heat and to fluff it up in the cold to retain heat. Hairless man does not have those options.

There is little doubt that persistent hunting has been used to advantage at times by humans. However, one must wonder if this activity has ever been prevalent in nature or whether it is a term invented to prop up evolution. It would seem that those who would propose it have never hunted. Once animals, say, antelope or deer, learn that man is a predator, they flee while he is still a mile away. Don't forget, men walk upright. Sneaking up on prey by crawling on all fours, to say nothing about the belly is not a good option. Take it from one who has done enough hunting to have tired it. It can be done but with probably no better results on the average than running down the prey. Lying in wait along a trail or near a water hole is the time tested means of successful hunting with or without weapons. It uses the least energy so when the prey arrives the hunter is rested and alert.

So, why do humans sweat so much? It is because they are the only species that does genuine work and work is totally associated with intelligence. We are not talking about the few days a year a bird spends building a nest or things like that. A man knows that he must work hard in the hot summer to build a dwelling for the cold winter. He must work to grow or gather food in times of plenty for the lean times. He works to make clothing and tools. He works to fortify his village from enemies both men and animals. It is man's lot to work and in the process to sweat.

Man works to improve his existance—something animals do not do. He works so he has more food, better food, a greater variety of food. He works so he has warmer clothing and a warmer dwelling where the roof does not leak so much. Animals do not like being out in the rain getting wet—it's miserable. If there is natural shelter at hand they will get out of the rain. Yes, even animals have enough sense to get in out of the rain. The author grew up on a farm. It was the boys chore on winter afternoons after school to spread out straw in the stalls for bedding and to put feed in the toughs. On cold, windy, snowy days the cattle would be gathered near the barn door with a couple of inches of snow on their backs waiting to be let in. They did not like being cold and wet any more than people did. It is just that without intelligence there was nothing they could do about it.

The fact that most people today would associate man's ability to sweat with endurance hunting is expected. Few modern people have ever really spent a day working where they sweated profusely. They only sweating they know about is while jogging so naturally they would associate sweating with running a marathon. However, on those hot days hauling hay on the farm or doing other labor intensive tasks we drank often to prevent dehydration. The endurance hunter would not have that luxury.

A few words have to be said about certain animal behaviors. For example, it is granted that beavers make lodges to dwell in especially in the winter. It is also true that they do that only in one instinctive way. They stay active but never overly so. We see squirrels scurrying around in the trees like they were working or at least on some specific mission. We see them about these activities except when we don't. There are long periods in the day when they do not do much. That is true of nearly all animals. Not so for humans, humans that have any concept of survival, that is. Welfare recipients in modern society do not count; they would be the first to go extinct. But, maybe not. They might be good survivors given the incentive.

Intelligence and Survival

Regardless of what a few evolutionists say, intelligence is generally considered to be a benefit when it comes to survival. However, it is assumed that human intelligence evolved along with his body. That, in turn, brings up another question that is seldom asked. If intelligence is a distinct benefit to survival, why aren't there animals that are a lot more intelligent then the ones we see? What happened to monkeys? Why are not chimpanzees a lot smarter then they are, or why aren't there other species between them and man? One would expect that there would be several competing species nearly as smart as we are. Please do not bother with Homo Erectus, Piltdown Man and some of those. When it comes to the missing links between men and monkeys there are simply too many frauds.

Any honest person must admit that chimpanzees are not intelligent. They are dumb animals. As shown below they are not tool users; they do not fabricate things; they do not think. Reasoning is unique to humans; it is categorically unique. From observation it should also be obvious that rationality did not evolve—a being has it or it does not. From that, one should conclude that a chimp is about as smart as an animal gets which leaves the problem of where man's body came from.

If rationality has no bearing on survival, then the animal that existed just before man got a rational soul was just as smart as we are now when it comes to survival. That would mean there would be no missing links in the fossil record because we are the missing links. That is, there would be no way to tell the difference between the animal and us. It would also mean the animal would have been a prolific tool user, and maker of clothing and dwellings, and not just by instinct like birds build nests, but to suit the environment.

If the human is not equipped to survive with an animal brain, just the opposite is true if he has a rational mind. The human body is perfect in that case. To start with being a biped and walking upright, he is much better equipped to see around him and hence locate food or dangers. Having fantastically agile hands with opposing thumb makes him a natural as a tool maker and tool user. And not needing his hands for locomotion permits him to carry things—tools, weapons or extra food. It is thought that perhaps the first invention of rational man was the bag, a natural extension of carrying things in his hands.

When a predator kills a large animal it eats its fill but must leave the rest. If it remains undisturbed the predator may return later for a second meal. But, all too frequently other predators will consume the remains as soon as the first leaves. Man, with his free hands, to say nothing of his bag, can carry extra food with him. Of course being intelligent and dexterous man can make clothing and shelter thereby extending his range of habitation.

What this all means is that getting the human body "evolved" to the point where it was when it was endowed with intelligence is not possible. That said, it is devilishly convenient to say that therefore this entire premise is incorrect and that intelligence evolved along with the human body. But, and a big but, there is a categorical or metaphysical difference between human rational thinking and animal intelligence. There have been hundreds of books written on this subject so no attempt will be made here to duplicate even part of that body of knowledge. Only a few common sense examples will be offered here.

One of the faculties of humans that is categorically unique to the species is that of speech. No animals can talk. They are at most given to a few simple instinctual utterances such as a cry of pain or a warning of danger. There is no hint of animals even beginning to develop speech. With language, even though primitive, man could communicate instructions to others or pass on learned techniques of the hunt or crafts for making tools. Wolves frequently hunt in packs which is an instinctual behavior that is always the same. Man can coordinate activities to suit the situation. Speech is the subject that is conveniently ignored, totally, by avowed evolutionists when they discuss evolution.

In this vein, scientists have supposedly taught chimpanzees in the lab to count. That is at most a learned reflex. It is about the same as teaching someone whose only language was English a few phrases of Russian to welcome a guest from Russia without a translator. However, both parties would know the English speaker did not know what he was saying. The Russian would accept it in good grace as a courteous attempt on the part of the English speaker to make him feel welcome. That the English speaker did not know the Russian language remained a fact.

Only rational beings are responsible for their actions. If you accept the existance of court rooms, judges, juries, and prisons, you accept that there is a categorical difference between a man who kills someone and a wolf that does the same thing. We do not have prisons for wolves.

In addition there are many other features that are odd for an animal like us that only make sense if we are rational. For example, we expend twenty percent of our energy on our brain which only makes sense if we are rational. Humans have perfect binocular vision and color vision—two things that fit perfectly with dexterous hands and rational thinking.

If one wants to stubbornly stick to the idea that intelligence evolved there is nothing that can be said. However, he must account for the fact that we have rational minds and can think, arrive at conclusions, make decisions, draw inferences, judge, conceive, discover and invent, and form logical sequences to go from what we do know to what we do not know. We are immanently qualified to survive in the world as it now exists and within a wide variation from the way it is now. These powers that man possesses simply could not have arisen out of rocks or interstellar gases. There is something beyond the matter and energy of the physical universe in the human soul. It could be stated a hundred different ways but it would not change the fact that the human soul is beyond the material universe.

Tool Using Animals

A few words must be said about "tool using animals." An article appeared in the journal *Anthropogi* that reports the results of an extensive literature search done by two researchers. They scoured the published literature for work done in the area of monkeys and their use of appliances which they differentiate as either *object* use or *tool* use. The difference is of questionable significance in the view of this author. The authors looked at monkeys in captivity, semi-free and in the wild. Cases where monkeys had been trained in any way to use objects were not included. First they define two distinct types of tool use.

> [T]ool use can be divided into two distinctive abilities. One, the motor act of manipulating an appliance or object, and two, the cognitive basis for understanding how the appliance functions (association *vs.* insight while solving tool use problems). [1]

Only then to they define what they mean by tool.

> In this regard . . . [is] a definition of tool use as the manipulation of an unattached environmental object, the tool (not part of the user's body), to alter more efficiently the form or position of a separate object, when the user holds or carries the tool *in toto* during or just prior to use and is responsible for the critical connection between tool and incentive. [2]

---

[1] Bernardo Urbani and Paul A. Garber, *Anthropogi*, XL/2, "A Stone In Their Hands... Are Monkeys Tool Users?," 2002, p. 183.
[2] Ibid.

That is to say, simply manipulating an object to affect some end constitutes object use, and only where there is evidence that the user understands the association between the object and its use is it tool use. In summary they say:

> The overwhelming number of published reports were associated with species of three primate genera, Cebus, Macaca and Papio. Moreover, the majority of observations described in the literature as "tool use" are examples of "object use." Considering the extremely small number of reported "tool-using" events and the fact that typically only 1 or 2 members of a group have been observed to manipulate objects as tools, we conclude, (1) the presently available evidence does not support the contention that monkeys naturally use tools, and (2) in those cases when monkeys are reported to use tools it remains unclear from the animals' behavior the degree to which they understand how the tool functioned (casual knowledge) in accomplishing the task. [3]

The above distinction between object and tool is a bit esoteric. At one place the authors give an example of the difference. They cite a case of the use of an object where a monkey hits a fruit on a rock to break it. But, if the monkey uses a rock to hit the fruit it is said to be the use of a tool. This author feels the concept of a tool should be more definite as in where the creature modifies a natural object so it better does a specific task. An example of this would be a fleshing tool. Sometimes it is difficult at first glance to see that a rock has been modified to be such a tool. This is because nearly any type of rock will do for the material as long as it fits reasonably well in the hand. Furthermore, there is no great precision needed in the modification. All that is needed is that the edge be reasonably sharp preferably with some amount of serration of the edge. It is usually the serration that will distinguish the rock as a tool.

------

[3] Ibid.

Chapter 7

What Philosophy Has to Say About Evolution

Various philosophical subjects will be considered in Part II. In particular self evident propositions will be treated in detail. But, we will make a start here since it has such a vital part in understanding why species did not come from rocks and gases.

Self Evident Propositions

This illustration might help people who are honestly trying to accept the "otherness" of the soul. One can set an empty quart jar on a table. Then place a cup of water next to it. So far so good. Now, pour the entire contents of the cup into the jar. What do you see? Most people would reply, water in the jar. But, more exactly what is before you. They would look puzzled and shrug their shoulders. Then you say what exactly you see is that the jar is not full. The other person would say sure, if what's the way you want to describe it, that's fine with me. But, you persist. What general conclusion can you draw from this experiment? Again nothing, they suspect they're being set up. Your reply, the general conclusion is that the greater, that is the volume of the jar, does not come from the lesser, namely the volume of the cup.

Your observer appears somewhat bewildered, and says something to the effect that this is all a game of semantics; that it is common sense. Yes, but, it is more. The statement, "The greater does not come from the lesser," is a philosophical concept called a self evident proposition, some times called an analytical proposition. There are a number of them and they are surprisingly important. A self evident proposition is a statement where if you know the definitions of the words you immediately see the truth of the proposition. Other common ones are "The whole is greater than any of its parts," and probably the most important one is "A thing cannot both be and not be at the same time." Our observer could say, "Tut, tut. I live a practical life and really do not care that half an apple is less than a whole apple—it's a given.

Where self evident propositions come into evolution is that the human intellect, the greater, does not, and cannot, come from the lesser, that is, an animal. It also says that life, the greater, cannot come from non-life, the lesser.

The only possible way around the above conclusion is in the belief in pantheism and that is an illusion. Pantheism says nature is god and god is nature. In that case there is something beyond pure matter, some mysterious force, in the sum total of all nature taken together that crates life and intelligence. That is what some scientists call god, a god that can cause the greater, that is, something that does not come from the material universe. There is no escaping that conclusion. The human mind can identify self evident propositions and they are invariant. The greater does not come from the lesser. It's common sense, the same as much of philosophy is common sense.

Recently the author has read a biography of Thomas A. Edison. Edison only peripherally connects with evolution, but it is instructive to see yet another angle. The biography was written during the last years of Edison's life so the author had access to the man and interviewed him at length, as well as people who had worked with him for most of his life. He also had complete access to Edison's notebooks and other documentation. This is mentioned here because we all have a mental image of Edison as the inventor of the electric light bulb, the phonograph and other things. He had hundreds of patents and . . . that's pretty much it. We know nothing about *him*, and just as importantly, his times. For example, it comes out in the book that he appeared at the prefect time in history where technology was fairly well advanced and yet it was still possible for one man, a loner, to make great inventions. Not only did he invent things, he designed and built the factories to manufacture his inventions. The man, William Shockley, who invented the transistor worked for Bell Labs. It is probably one of the greatest inventions in history, but he certainly did not build Bell Labs from scratch, nor did he design and build the factories that produced transistors.

The point is that most people have about the same idea about Darwin and evolution as they do of Edison, both are vague presences. Evolution is taken for granted today and thus is uncritically accepted by nearly everyone. When you start to look closely at evolution, at the details, at where it came from, the whole thing looks pretty shaky.

True Origin of Species

Having discussed in many ways why species did not happen by happy accidents, how does one account for the fact that they exist? There is a book called *Nature's Causes* by Richard J. Connell. In it he has a twenty-eight page chapter devoted entirely to evolution. He accepts that evolution took place in many small steps or mutations. And he allows that the fossil record essentially demonstrates, not proves, evolution. This author would not be that generous. But, it does not matter to him because he takes the work of Aristotle and others such as Thomas Aquinas and updates them to the fossil record, DNA, etc., that we have today. Here is a snapshot of the line of reasoning quoted from Chapter 10.

> Here, however, the question is about an agent cause responsible for the *origin of a new kind of operational capacity that determines a new kind of substance* [that is, a new species.] (emphasis his.) [1]

Then he goes into greater detail of what he means.

Because facts are immune to attack, we must concede that the species came into existence successively and we also must concede very probably they originated with some dependence of the posterior species on the prior. Given such concessions along with the axiom which governs agent causes we must say that the *anterior species cannot play a role greater than that of instrumental agent,* which means that besides the progenitor species, we must infer— not postulate or hypothesize—the existence of another principal causal action that is proportioned to the actual effect, the nature of the new species. And since the principal causal action is not that of another organism, of a material agent discernable directly through experience, we must also conclude that the principal agent is extrinsic to the realm of nature, much as man is extrinsic to the realm

---

[1] Richard J. Connell, *Nature's Causes*, New York, Peter Lang Publishing, 1995, p. 164.

of the artifacts that he produces. This agent we shall call a "universal" rather than an "univocal" cause. (Univocal causes produce effects that are the same species as themselves, for example organisms, whereas a universal cause of the sort we have just inferred is the agent that produces the first individual of a new natural species that is manifestly not simply a reproduction of the cause itself.) To repeat: we have argued to, but not postulated or hypothesized, the existence of an agent responsible for the collocation of operational powers and the consequent proportioning of the species to its environment, an agent that is extrinsic to the physical world. Our argument resembles the one we use when from the fluctuation of the compass needle we conclude to the existence of a magnetic field; in essence our argument to a cause outside nature is not different. (Emphasis his.) [2]

He goes on to explain that this is a somewhat new way of looking at the origin of species as opposed to evolution or creation.

We wish to emphasize that our position is not that of a creationist as that term is ordinarily used, and the reason is plain. The creationist puts the divine causality on the same plane as the theory of evolution, which means that he takes it as an hypothesis from which the natural regularities are to be deduced or explained; but that is not the way the Cause of Creation enters our understanding of natural things and events. For instance, the theory of evolution attempts to account for why certain species appeared at such and such a time under such and such conditions, and despite the difficulties it aims to account for the particulars of the evolution of species. In contrast if one starts with the proposition that God is the cause of species, then this argument is apriori and he cannot account for the first appearance of species; he cannot in any way tell us why God created these species at this time under these conditions, nor is that the purpose of Genesis. In sum, regarding the specifics of the origin of species we cannot know the divine mind. Nor have we here, in this work invoked the divine causality to explain the origin of species, as the reader has seen. Our procedure started from an effect—the first appearance of an individual of a new biological species—and from that effect conclude to the exis-

---

[2] Ibid., p. 165.

tence of an agent extrinsic to nature. We argued from nature to a cause outside nature, not from a cause outside nature to nature herself. [3]

The point of the above quotes is to show that an agent that is extrinsic to the physical world is needed to make a new species and, through the principle of causality, this eventually arrives at God. Therefore, it is the wisdom of the ages, starting with Aristotle, who predated Christianity, that every new species requires an agent outside of nature to make it. This was arrived at and sustained by the greatest minds the world has ever known. They reasoned meticulously through many steps carefully defining each term in the argument. On another plane, this is not really such rocket science. In simple terms, what they are saying is that, "There is no such thing as a free lunch;" "The greater does not come from the lesser;" "You do not get something for nothing." Anybody can understand that.

One of the terms used in the quote above that is not defined is instrumental agent. A simple definition will be repeated here.

But, knives and chisels are used by sculptors to carve statues, and when we compare these tools to the production of a statue of Socrates, say, we see that they are not sufficient by themselves to bring about the result. Yet in the hands of sculptors, tools serve to produce effects that exceed the capacities of the tools themselves. The sculptor acts as the principal agent, the principal active cause, while the tool acts as *an instrumental agent or cause*; yet it does stand in a real causal relation to the effect. A tool "produces" only in a diminished sense, and its own proper action is elevated by the principal agent. (Emphasis his.) [4]

In a certain sense this closes the book on evolution, not that there is not still a lot to be said about it. If the reader wants to believe in small step by step changes in organisms, that's up to him. But, when it comes to the step to a new species, it requires something outside of nature. Connell quotes Aristotle and St. Thomas in Chapter 10 so the reader could see the original text on the various ideas, but they are hard to un-

---

[3] Ibid., p. 166.

[4] Richard J. Connell, *Substance and Modern Science*, Notre Dame, Indiana, University of Notre Dame Press, 1988, p. 161.

derstand. For example, he says in the introduction that unless you have read his *Substance and Modern Science*, you will not understand substance and that organisms are substances rather than machines or mere organizations of chemicals.

This serves as a point to introduce one of several reasons why one can make the statement that if man ever travels to the stars, it will be Catholics that make it possible, not that everyone making the trip will necessarily be a Catholic. This reason is the idea of substance. Through the Holy Eucharist by means of Transubstantiation, Catholics are well aware of, and accept, the idea of substance because in this case it is the substance that changes and the properties or accidents do not. As a result, Catholics accept that substance is the basis of reality. No other religion puts substance in such a preeminent position, or even mentions it. One might suspect that most scientists do not accept the idea of substance and hence reject Aristotle and St. Thomas. That will eventually limit what they can do. Science in general that is not aided and guided by philosophy will dead-end and become a cult, not unlike evolution has become.

Chapter 8

Science and Evolution

We start this chapter by discussing a book where the author tires to make a case for two evolutions, one a philosophy and one a science. Then the "science" of the fossil record is discussed. This is followed by a treatment of a bed rock theory of modern physics that turns out to be little more than another attempt to do away with God.

Chance or Purpose

This small subsection is added as an example of how people's minds have been shaped by evolution. The title is also the title of a book *Chance or Purpose* written by Christoph Cardinal Schönborn. His intent was to give a definitive treatment of evolution giving due place to creation and science. He states many times through the work that there are two concepts of evolution at play in modern thinking. One he calls the ideological concept of evolutionism and the other the scientific theory of evolution. He offers the following with regard to evolutionism.

> Belief in creation lays the foundation for the basically positive view of creation and of all its manifestations. . . . Evolutionism as a way of seeing the world (not as a scientific theory) has far greater difficulty with this. For this worldview, there are not really any species, for things have no existance of their own. What we regard as "species" are in fact merely "snapshots" in the great stream of evolution. Everything is just transition and a stage being passed through, and each individual is merely a fluke, which had the luck to survive because it was "more fit" than the others. This is

certainly a short-sighted view of the variety of creation. The way men marvel at the variety of nature gives us a hint of something different. Above all, it seems to me, evolutionism as a worldview cannot actually offer any reason why anything has any value in itself, if everything is, so to say, merely a transitory stage in the stream of evolution. [1]

The opposite of evolutionism is creation. On this subject he has this to say.

> Are there then, "individual acts of creation" after all? Yet how could we establish their existence? Here we need to refer to a perfectly simple distinction that people prefer to overlook. This is the distinction between a precondition and a cause. In order for life to come into being on our planet, a whole series of preconditions were needed, without which there would be no life. Yet these preconditions were—and are—only the framework or conditions for life to come into being. They do not constitute the creative cause of life. They all play a part in life's coming to be; yet the new element in the development of the world, which we call life, cannot be derived from them. For it to come about, it truly needs the creative act of God, the "divine spark", in order to come into being. . . . The great "leaps" by which the stages of evolution ascended each had their necessary preconditions, which cannot however be the things that created those realities. They are genuine contributory causes, but not the actual creative cause. [2]

That leaves the scientific theory of evolution to discuss. It is significant that Schönborn never adequately defines what he means by this. It is as if he assumes everyone knows exactly what he means. Or, perhaps it is left open so his readers can assign any and all ideas they desire to it. He comes the closest to science in the summary to his book there he makes the following comment.

> The correct path to follow is not to choose between the "Darwinian story" and "creationism", as people like to do, but a co-

---

[1] Christoph Cardinal Schönborn, *Chance or Purpose*, San Francisco, Ignatius Press, 2007, p. 60.
[2] Ibid., pp. 82-3.

existence of "Darwin's ladder" and "Jacob's ladder." There is a great deal to support the view that life has developed through a long process, a gradual ascent from the most simple beginnings up to the complexity of man. It is marvelous to penetrate ever more deeply into the common building blocks of life and thereby into the way that all life is related. [3]

However, having accepted that each new species requires not only the proper preconditions but also a specific creative act, he essentially does away with evolution whether ideological or scientific. The only science left is to categorize the existing species into a family tree of living things and by extension the various fossils, though in this latter case most parts of the tree are missing.

This brief review of *Chance or Purpose* was done to show how ingrained is the idea of evolution. Schönborn is a member of what could be called the intellectual elite, that is, a certified smart person. He has written many books and is well respected. However, if he were to say out right that there is no science to evolution he would immediately find he was no longer so elite nor certified. The truth of the matter is that he is caught as are most people not being convinced one way or the other. If he takes the logic as presented above step by step he sees there can be no evolution, but in his gut he cannot let it go.

The Fossil Record

The fossil record has been mentioned a few times but little has been said about it. The main reason is that there is not much that can be said. As a matter of fact, serious scientists are backing away from it.

The missing link between man and the apes . . . is merely the most glamorous of a whole hierarchy of phantom creatures. In the fossil record, missing links are the rule. . . . The more scientists have searched for the transitional forms between species, the more they have been frustrated. [4]

Adler goes on to say:

---

[3] Ibid., p. 169.
[4] Jerry Adler, *Is Man a Subtle Accident?* Newsweek Magazine, November 3, 1980.

Seventy years after quantum theory revolutionized physics, an oddly analogous change has occurred in the theory of evolution and it is just beginning to filter down to public understanding. Evidence from fossils now points overwhelmingly away from the classical Darwinism which most Americans learned in high school: that new species evolve out of existing ones by the gradual accumulation of small changes, each of which helps the organism survive and compete in the environment. Increasingly, scientists now believe that species change little for millions of years and then evolve quickly, in a kind of quantum leap—not necessarily in a direction that represents an obvious improvement in fitness. The theory is still being worked out. Among other points of contention, it is uncertain whether the leap takes place in a few generations or over tens of thousands of years. But at a conference in mid-October at Chicago's Field Museum of Natural History, the majority of 160 of the world's top paleontologists, anatomists, evolutionary geneticists and developmental biologists supported some form of this theory of "punctuated equilibria." [5]

Commenting further on the conference mentioned above J. W. G. Johnson has this to say:

Having admitted that the evidence for evolution is missing, these experts are now working on a theory of evolution without evidence. They are suggesting evolution by huge jumps which would leave no fossil evidence. [6]

The book *Evolution?* is one of many good works that takes to task all of the tenets of evolution and demonstrates by presently known facts that they are false.

Punctuated equilibrium has as its basis the unspoken assumption that there must be something greater than simple nature to produce new species as was demonstrated earlier. In some aspects it is similar to the idea of Intelligent Design that was also mentioned earlier. That theory also forces a person to accept some force beyond nature to account for the variety of species without mentioning God explicitly. However, it has a

---

[5] Ibid.

[6] J. W. G. Johnson, *Evolution?*, Australia, 1982, p. 13.

down side in that many people who believe in evolution also believe in pantheism whether they admit it or not. Pan is the Greek god of field and forest. He is most commonly depicted as a goat walking on his hind hoofs while paying a flute. Pantheism says that nature is god, and conversely that god is nature. That being the case there is no limit to what a cloud of interstellar gas, or a pile of rocks can do—they are a god.

Modern Physics

It is time to take another step, not forward but to the side, in our study of the modern mindset. The previous pages have been dedicated to the study of the oddly important subject of evolution and how it has replaced our traditional way of thinking with a way of not thinking. This section is added to show how modern physics has its own theory equivalent to evolution.

They both start form the point made in the introduction of this book about how there should be no distinction between science and philosophy since both are exercises in the search of knowledge. And as has been suggested the reason why scientists reject philosophy as having anything to do with science is that one branch of philosophy is theology and they stand firm on the position that religion and science do not mix.

With that as a irreconcilable premise, all of the sciences tend to go off the rails in odd ways. Certainly the life sciences would be different if evolution were not a point of "rock hard fact" that had to be included in any and all science dealing with life. That same stumbling block is seen in physics and it is worth our while to do a brief study of one aspect of that.

What is called modern physics deals with such things as subatomic particles, quantum mechanics, and Einstein's theories of relativity. Here it is Einstein that will be our point of interest. It starts with what is called the Michelson-Morley experiment conducted in 1887. The reason for that experiment was to prove the theory that all space was filled with a substance called ether. The reason ether was postulated was from the discovery that light was a wave. Sound waves travel through a medium like air, water or solid objects. No medium, no sound. There is no sound in outer space. It seemed only reasonable that there would have to be a medium for light waves to travel in, hence ether was proposed.

The particular experiment was to take a beam of monochromic yellow light and split it into two parts. One part would be made to travel against an ether wind and the other would travel perpendicular to the wind. The wind was to be created by the motion of the earth traveling around the sun. The two beams of light would be reflected back and then be recombined. The part of the beam headed into the wind would be retarded slightly with respect to the other and this would show up as an interference fringe in the interferometer the investigators used.

To understand the idea used in the experiment it is good to take a simple example. Assume there is a river that has a uniform current that flows at ten miles an hour. Then assume a traveler wants to go up stream ten miles in a boat that travels twenty miles per hour in the water, whereupon he immediately returns down stream to the starting point. The up stream trip takes an hour because the speed of the boat is retarded by the stream so its speed is only ten miles per hour with respect to the bank. Likewise the return trip sees the boat going thirty mph which is the combined speed of the boat and the current. This part of the trip takes only twenty minutes meaning the entire trip of twenty miles takes one hour and twenty minutes. Now assume the same traveler wants to go across a calm lake with no current a distance of ten miles and then back again. Once again the boat can go twenty mph so the trip across the lake takes a half hour. The same is true of the return trip for a total time for the twenty mile trip of one hour. Notice how the round trip up the river and back takes longer than in calm water. That is the idea used in the Michelson-Morley experiment.

The actual experiment with light waves would have the part of the perpendicular beams going across the current but the speed of the light wave relative to the speed of the earth is so great that the "cross wind" would not be a factor. For reference the speed of light is 186,000 miles per second. The speed of the earth around the sun is about 18.5 miles per second. Through numerous other tests it was known that the apparatus could easily detect an ether wind caused by the speed of the earth around the sun. They expected to see a difference in the two light beams of 0.4 wave length while the interferometer they used could detect a difference as small as 0.01 wavelength, one-fortieth as much as the expected result.

The Michelson-Morley experimental apparatus was constructed with meticulous care since the experiment had been tried before and the results were inconclusive. The experiment of 1887 had the whole apparatus floating in a pool of mercury. The idea was to make a measurement with one axis pointing into the direction of motion and then rotate it ninety degrees thus eliminating most sources of error. The moment of truth arrived and the experiment was conducted, and conducted, and conducted.

All tests produced a null result, that is, they could not detect the ether wind.

This led to great consternation. Other than possible faulty experimental method which all reasonable reviewers had agreed had been eliminated there were two main explanations for the results. The first was that there was no ether. If that solution were accepted it would mean a re-thinking of the physics of light and associated physical phenomena. The other was even worse. It said that there was an ether but that the earth was not moving through it which in turn meant that the earth was the center of the universe. That led directly to the position that the narrative of the creation of the universe and man as stated in Genesis was correct. That idea was out of hand rejected.

Through the years a number of researchers recreated the experiment under somewhat different circumstances, all of which led to the same results. Also, theoretical physicists were casting about for some way to explain the apparent lack of ether. In addition other experiments such as a telescope filled with water, which will not be gone into here, were giving the same results and implying the same conclusions.

As an aside, this is not unlike the biologists of Darwin's time casting about for ways to explain life and the various species without resorting to creation. And do not miss the point that the Michelson Morley experiment was conducted in 1887 just 28 years after Darwin's first edition appeared.

Eventually in 1905 Albert Einstein published his famous special theory of relativity which, among other things, did away with ether. "He did so, by his own admission, in order to have an answer for the 1887 Michelson-Morley experiment which showed the Earth was motionless in space."[7]

Special relativity did not include gravity so in 1915 Einstein came out with his general theory of relativity which did. To make things work he was forced to bring back ether in very limited ways. In these theories he postulated that the speed of light was a constant regardless of the speed of the observer. His equations led to rather unorthodox conclusions such as when an object traveled at a speed approaching the speed of light the speed of time for the object approached zero, the mass of the object approached infinity, and the length of the object in the direction of travel approached zero. That meant that it was impossible for anything to reach the speed of light because the mass of an object could never become in-

---

[7] Ibid. Sungenis, Vol. I, p. 199.

finite.

At the time nobody was particularly worried about actually trying to approach the speed of light. They were instead interested in the meaning of these equations with respect to an object moving 18.5 miles per second, the speed of the earth around the sun. In that case, the Michelson-Morley experiment would naturally produce a null result. They looked at it this way. Einstein's equations said an object was shortened in the direction of motion proportional to its speed relative to that of light. That meant that the apparatus used by Michelson and Morley had been shortened in the dimension headed into the ether wind just enough to give a null result. It did not matter if you rotated the apparatus in the pool of mercury because whichever dimension faced the direction of travel would be shortened when the other returned to its "normal' length as it turned perpendicular to the line of travel. And, do not bother trying to measure how much it had been shortened because your measuring stick would be shortened by exactly the same amount.

Presto, Einstein solved the ether wind problem without having to resort to creation. From that time on the theories of relativity were accepted and taught as "gospel" to every physics student, not that many physics students actually understood all of the mathematics involved in the theories. Einstein deliberately made the mathematics more difficult that was needed so not many people would adequately understand the theories enough to pose objections. That was somewhat similar to what Isaac Newton had done centuries before in his presentation of the law of gravity as was presented above.

Before we leave this subject it must be observed that it is not an unsupportable idea that the earth is the center of the universe, is not moving around the sun and indeed is not even rotating on its axis. If that were the case and the sun revolved around the earth while all the other planets still revolved around the sun everything still works. It is one of the fundamental tenants of relativity that everything is seen *relative* to the position of the observer, that is to say, where one puts the center of his coordinate system. That in turn means that geocentrism, the theory that the earth is the center of the universe and not rotating, makes perfect sense. Even the famous theoretical physicist, Steven Hawkins, said geocentrism was fine with him. It only made some of the math more difficult.

If the reader is interested in the intriguing subject of geocentrism he will find that there has been a good deal of serious work done on the subject. The reference given above, Robert A. Sungenis and Robert J.

Bennett, *Galileo was Wrong*, 10th edition is one of the best. In three volumes totaling two thousand pages Messrs. Sungenis and Bennett do a masterful job of demolishing the theories of relativity along with many other long held theories in physics. The reader will at first ridicule the idea as did this writer. But, by the time he is through the massive tome it is clear that at least there has been a lot of misleading information used as fact to bolster careers of scientists during the last hundred and fifty years. What one finds is that research in the last fifty years has shown that much about Einstein's theories is wrong. This in turn, has led to something not unlike a cult concerning Einstein where his followers put patch upon patch to keep relativity from falling by the way. When perusing reviews of Sungenis' work on the Internet it is obvious that those who have read it agree with its findings and those who have not read it are vociferously against it.

Not to belabor the point but it is important for the reader to see the significance of relativity. The reason for its name is it hinges on the idea that all physical measurements are relative to the point of view of the observer. Relativity supposedly got the scientists off the horns of the dilemma of either there was no ether or the earth was not moving. But, by its definition it answered perfectly the premise that the earth was not moving just as the Michelson-Morley experiment showed. From that reasoning there was no need for the complex, and somewhat outlandish, theory that just happened to have as one of its tenants that the apparatus was shortened in the line of travel by exactly the amount needed to produce a null result. Outlandish or not, relativity was taught with a vengeance when this writer was in college in the 1960s. The mere mention of ether brought scoffs from the professor. Exactly as with evolution in the life sciences, no student would advance to the Ph.D. level in physics without bowing to the sacred cow theory of relativity in that field of study. Today, there is a slow movement away from at least some of the claims of relativity. In fact, there is now being postulated the idea of dark energy and dark matter in the universe. The latter could be the "missing" ether of yore.

PART II

THE ESSENCE OF TIME

Chapter 9

Time According to Aristotle, St. Thomas and St. Augustine

Introduction

This treatise on time is meant to be an adjunct to the study of evolution because evolution is as much about time as it is about the changes in organic structures. That said, the subject of time does not seem to be of interest to most people even evolutionists. Yet that should not be too surprising because most evolutionists are not much interested in evolution either other than to be sure everyone else believes as they do. However, a study of time could either bolster the case for evolution or refute it.

Here, we will look at time from several aspects. It is necessary to go far afield so the reader becomes conversant in the meaning of time from the philosophy of it, the physics of it and even the theology of it. Some things discussed may seen irrelevant but taken as a whole it is hoped that the reader will absorb an appreciation of a subject that is rarely mentioned.

Essence

Since this is to be a study of the essence of time it is well to start with a few words about the meaning of essence. In this regard there are two related concepts that must be distinguished—essence and being. On a fundamental level one can say that "the essence is what is signified through the definition of a thing." [1] And a being is an exemplar of that essence. In the real world of composite substances like rocks and people a composite substance is composed of matter and from. In people the soul is what forms the primary matter into us. "Matter in this context is

---

[1] Thomas Aquinas, *Thomas Aquinas On Being and Essence*, translated by Armand Maurer, 2nd Edition, 1968, The Pontifical Institute of Mediaeval Studies, Toronto, Canada, Chapter 2, § 1, p 34.

84

the primary matter that combines with substantial form to constitute a sensible substance. In itself it in not actually something or knowable. It is the pure possibility of receiving from." [2] In the case of substances "essence embraces both matter and form." [3]

However, time is not a composite substance of the type stated so we must look further. There is a category of essences that do not contain matter and these are spirits, of which more will be said later. These come in three types, the infinite spirit, namely God, angels, and human souls. Time certainly doesn't fit into any of these three categories except perhaps God since he is "all in all." That leaves the idea of accidents as essences.

"Because essence is what the definition signifies, as has been said, accidents must have an essence in the same way that they have a definition. Now their definition is incomplete, because they cannot be defined without including a subject in their definition." [4] Or more succinctly, "An accident is an essence whose nature it is to exist in a substance as in a subject." [5]

That leaves the interesting prospect that the essence of time is an accident of human beings that are alive. If it can be shown that time only exists for humans that are alive that would be a reasonable statement. In what follows an effort will be made to show that statement is true.

Aristotle and St. Thomas Aquinas

Normally one would start this study with Aristotle. Here St. Thomas Aquinas has been included because his thinking on time follows almost exactly that of Aristotle. Without having to get too involved in either Aristotle or St. Thomas use will start by making use of a book by H.D. Gardeil where his section on time has many references to both philosophers. In his section on *The Nature of Time*, Gardeil says:

But, though time is not motion, it is nevertheless unseverable from motion. Take away all change or motion, and time disappears. That is why the awareness of time dies with the awareness of change, as happens in sound sleep. No motion, no time, so much is true. But, motion is time, no. Hence, though not identical with it,

---

[2] Ibid.

[3] Ibid., p 35.

[4] Ibid., p 66.

[5] Ibid.

time is yet somehow affiliated with motion. What is the affiliation? What, in other words, is time?

Aristotle's answer is progressive. Time, he says, is continuous; it attends motion, and motion implies extension, which is continuous. This, then, is one thing that defines time, it is continuous. Secondly, in magnitude there is a before and after, namely, of position. . . . But, note, thirdly, what we do when we perceive a before and after. We distinguish phases of the motion, marking off, mentally, one part from another. That is, we perceive the motion as measurable and numerable, and number it. To differentiate within a quantity or magnitude is equivalent to numbering. In general, therefore, we may say that motion plus numbering equals time a thought that St. Thomas sets forth as follows; "Since succession is found in all motion, and one part follows another, in numbering the before and after of motion we apprehend time, which is nothing else than the number of before and after in motion." [6]

Here we have a definition of time but a definition does not always tell us what a thing really is. In the following section on *The Reality of Time* Gardeil attempts an answer.

To know the definition of time is one thing, to know what sort of reality it is may be quite another. So evanescent is time that the question may well be asked with Aristotle whether it has any objective existance at all. Can a thing be real if its parts do not really exist? Yet, time appears to be made of parts that to not exist. It has past and future, but the past is no more, and future is not yet. True, there is also the present moment, but that alone, whatever its actuality, does not constitute time. Add to this that time, it seems, can hardly exist without a mind to piece the parts together. Time is the number of motion. But, without something that can count there should be no number. Yet only an intellect can number. It seems, then, that without a soul (in the sense of intellect) there could not be number, hence no time. [7]

---

[6] H.D. Gardeil, O. P. *Introduction to the Philosophy of St. Thomas Aquinas*, Vol. II, Cosmology, B. Herder Book Co. English translation 1958, pp. 121-2
[7] Ibid. p. 123

From this one would conclude that time is strictly a subjective thing. That is not strictly true as we see further on in Gardeil as he continues to make use of Aristotle and St. Thomas.

Aristotle allows that time in its full meaning cannot exist apart from mind. . . . Nevertheless, time is not a sheer subjectivity. The mental work of discriminating before and after and relating them to each other has an objective foundation, being grounded in the motion of which before and after are parts. Granted that motion is an imperfect reality. . .it is nevertheless a reality. Thus, speaking of the objective and subjective elements of time, St. Thomas writes: "That part of time which is as it were its *material* element, namely, the before and after, is founded in motion; but the *formal* element is completed in the soul's activity of numbering. And that is why the Philosopher says that without a soul there would be no time." Emphasis added. [8]

Here St. Thomas makes time a substance by saying it is composed of matter and form. We will have to look at that more closely because the material of time being made of the before and after is made of nothing because the past is no more and the future is not yet. And, he leaves out the present entirely.

Another modern author is Paul Glenn who has the following to say about time in which he also relies on both Aristotle and St. Thomas Aquinas. "Time is an accident (like position, or action) which determines a reality in its position with reference to before and after. Examples: at midday, this evening, at five o'clock, next Tuesday, in 1492, before midnight, after supper." [9]

He goes more in-depth later as he says:

Bodies with quantity are subject to *change*. Change is movement or motion, for "change is a transit, a going-over, a movement from one state of being to another." Now, movement or motion is a matter of "now this—then that"; it is a matter of "before and after." And motion or change, under the aspect of before-and-after, is the

---

[8] Ibid. pp. 123-4.
[9] Paul J. Glenn, *An Introduction To Philosophy*, St. Louis, Mo., B. Herder Book Company, 1943, p. 249.

basis of *real time*. Time in itself is <u>described</u> as a continuous and numerable series of motions under the aspect of before-and-after. Man <u>conceives</u> of time as a *measure*, just as he conceives of space as a *container*. But just as space in its reality is the real extension of bodies, so time in its reality is the continuous numerable succession of bodily movements (of sun, of stars, of moon). Time *as a measure* <u>is</u> logical being, [10] not real being. . . . Besides real time we have *ideal time* which is the mind's concept of all possible numerable and continuous movement; and we have *imaginary time* which is the fanciful envisioning of real time indefinitely extended. Real time is necessarily finite, for it is finite motion in a finite world of finite bodies. Ideal time and imaginary time are *indefinite* or *potentially infinite*, but never actually infinite. Thoughtless people sometimes confuse ideal or imaginary time with *eternity*. But eternity is, strictly speaking, the opposite of time. It is an endless "now"; it has nothing of "before and after" which is the essence of time. (All emphasis his except underlines which are mine). [11]

Notice that in speaking of eternity above Glenn defines eternity as: "It is an endless 'now'; it has nothing of 'before and after' *which is the essence of time.*" Emphasis added. In so doing he defines the essence of time as "before and after." Above he says time is an accident so he seems to be saying that time is an accidental essence that only exists in an intellect.

There are some additional things one finds interesting about Glenn's treatment of time. The first is that it is all in terms of "before and after," since he never mentions the present. Aristotle, St. Augustine and St. Thomas dwelt on before and after, too, but also considered the present. The closest Glenn comes to mentioning the present is in the definition of eternity which is and endless "now."

The second point is that he uses "numerable" to describe time. This again is in accordance with the others mentioned. From this we could say that time is the number of motion, according to a before and after. But, that describes what we *do* with time not what time *is*. It is like asking

---

[10] Logical being is being of reason: that which can only exist in the intellect conceiving it. H. D. Gardeil, O. P., *Introduction to the Philosophy of St. Thomas Aquinas*, Vol. IV, Metaphysics, B. Herder Book Co., 1967, pp. 299.

[11] Ibid., Glenn, pp. 263-4.

what air is and someone saying that air is the gas we breath in and out to stay alive. That statement tells want we *do* with air not what air *is*.

Gardeil refers briefly to the present as follows: "Can a thing be real if its parts do not really exist? Yet time appears to be made of parts that do not exist. It has past and future; but the past is no more and the future is not yet. True, there is also the present moment, but this alone, whatever its actuality, does not constitute time." [12]

One could argue that the present is the only time that does exist because as they all agree, the past and future do not exist. Since we do exist we exist only in the present so that part of time *must* exist and represents the only real part of time. As Glenn says above, "Time is an accident (like position, or action)." Whether or not time is an accident is the question before us. But, more importantly he associates time with "position" as in space or "action" which is motion which in turn is the change of position in space, and time is used to determine motion. That means he is using time to define time. In any case, all of this measuring of before and after only describes what time is used for not what it is.

Perhaps what the philosophers do not make clear is they are looking at time as an accident. In addition to position and action, color, for example, is an accident. All of these accidents are part of physical beings. Attaching time to a "thing" as an accident is like attaching the thing to space. Of course, a "thing" needs space to exist just as it needs time. But, we do not think of space as being an accident because all things need space and accidents are what distinguish one "thing" from another. In like manner, all things need time.

An accident does not exist by itself but only in a thing. There is no such thing as yellow. It must be a yellow flower or a yellow button. However, yellow does exist in our minds apart from any specific thing. And, one could say that time does exist in our minds apart form any specific event. The difference is that the only place that time really does exist is in our minds since the future is not yet, the present has no length and the past is no more.

St. Augustine

To take this a step further, we are indebted to St. Augustine for one of the best discussions of time in his *The Confessions of St. Augustine*. He lived from A.D. 345 to 430 and wrote this work in the years around 400

---

[12] Ibid., Gardeil, Vol. II Cosmology, p. 123

so he was not yet fifty. In Book 11 of that work, he is pleading with God to give him insight to this vexing problem of time. He considers the question of the beginning of the world and with it the beginning of time and shows that time and creation were made at the same time. But what is time? To this Augustine devotes a insightful analysis of the subjectivity of time and the relation of all temporal process to the abiding eternity of God. Chapters I through XIII of Book 11 are not particularly relevant to our purposes here, though the reader is encouraged to read them. As a result we start with Chapter XIV. Here we are using a translation by Albert C. Outler, Ph.D., D.D. It uses contemporary language and it is well referenced. As one might expect there are many translations of such a famous work which was originally written in Latin. As such there are times where the interpretations vary.

To summarize, then, though inadequately, St. Augustine looks at time future, present, and past arriving at the conclusion that the future does not exist because it is not yet, the present does not exist because it has no length, and the past does not exist because it is no longer. Finally he arrives at the point that the only place time, especially future and past, exist is in the human mind. His conclusions are the same as those of Aristotle and St. Thomas. It is not clear if he had access to Aristotle's work on time. Of course, St. Thomas was still centuries in the future when St. Augustine lived. At times his treatment is perceptive, at other places he stops and appeals to God to give him the light to understand his topic.

To give the reader a taste of the work Chapter XIV will be reproduced here followed by parts of other chapters.

Chapter XIV

There was no time, therefore, when thou hadst not made anything, because thou hadst made time itself. And there are no times that are coeternal with thee, because thou dost abide forever; but if times should abide, they would not be times.

For what is time? Who can easily and briefly explain it? Who can even comprehend it in thought or put the answer into words? Yet is it not true that in conversation we refer to nothing more familiarly or knowingly than time? And surely we understand it when we speak of it; we understand it also when we hear another speak of it.

What, then, is time? If no one asks me, I know what it is. If I wish to explain it to him who asks me, I do not know. Yet I say with confidence that I know that if nothing passed away, there would be no past time; and if nothing were still coming, there would be no future time; and if there were nothing now, there would be no present time. But, then, how is it that there are the two times, past and future, when even the past is now no longer and the future is now not yet? But if the present were always present, and did not pass into past time, it obviously would not be time but eternity. If, then, time present—if it be time—comes into existence only because it passes into time past, how can we say that even this is, since the cause of its being is that it will cease to be? Thus, can we not truly say that time is only as it tends toward nonbeing? [4]

Chapter XV

Thus it comes out that time present, which we found was the only time that could be called "long," has been cut down to the space of scarcely a single day. . . .And that one hour itself passes away in fleeting fractions. The part of it that has fled is past; what remains is still future. If any fraction of time be conceived that cannot now be divided even into the most minute momentary point, this alone is what we may call time present. But this flies so rapidly from future to past that it cannot be extended by any delay. For if it is extended, it is then divided into past and future. *But the present has no extension whatever*. Emphasis added.

Chapter XVIII

Give me leave, O Lord, to seek still further. O my Hope, let not my purpose be confounded. For if there are times past and future, I wish to know where they are. But if I have not yet succeeded in this, I still know that wherever they are, they are not there as future or past, but as present. For if they are there as future, they are there as "not yet"; if they are there as past, they are there as "no longer." Wherever they are and whatever they are they exist therefore only as present.

Chapter XX

But even now it is manifest and clear that there are neither times

---

[4] Albert C. Outler, Ph.D., D.D., *The Confessions of St. Augustine*, Philadelphia, Westminster Press, 1955.

future nor times past. Thus it is not properly said that there are three times, past, present, and future. Perhaps it might be said rightly that there are three times: a time present of things past; a time present of things present; and a time present of things future. For these three do coexist somehow in the soul, for otherwise I could not see them.

Chapter XXVIII

But how is the future diminished or consumed when it does not yet exist? Or how does the past, which exists no longer, increase, unless it is that in the mind in which all this happens there are three functions? For the mind expects, it attends, and it remembers; so that what it expects passes into what it remembers by way of what it attends to. . . . [5]

---

[5] Ibid.

Chapter 10

Other Thoughts About Time

The Human Meaning of time

To be human is to have a deep sense that something is missing, that we are lacking something. This occurs on a spiritual level that can be seen in two forms. One might call the first a direct longing which was probably said best by St. Augustine, "Our souls are restless, Lord, until they rest in Thee." The second is a response to an emptiness that we can, at least partially, satisfy by having a connection with the past and the future. This is commonly done by forming families, though that is not the only way. Certainly it is not proposed that it is necessary for every man and woman to marry and have children, just that societal structures must be oriented in a direction that those who do have families have the support of the wider community to be successful.

Through fertility time becomes a continuity as each person sees himself as having come from the previous generation, his parents, and children as connecting him to the future. This is why we find a need to connect with our past, why we need a history. Beings that do not have this basic, supernatural need do not have a history. Animals are fulfilled in that they know what to do as dictated by instinct. Humans do not know what to do and must be taught. And what must we be taught? It is how to do the things our ancestors learned that will help us survive; it is the passing along of culture, of making monuments, of learning our history. This man of time, of history, embodies in himself his tribal past. He has an impetus to create and sustain institutions that best permits heirs. In the modern, liberal culture our history is being rewritten, lied about, torn down and denied; our monuments are being destroyed.

We see it in the destruction of the Civil War memorials in the South. We see this in the way schools denigrate the founding fathers of our country. We have even seen it as the liberals have taken it upon themselves to rename one of the lakes in Minneapolis—Lake Calhoun. They have replace it with Bda Maka Sha meaning Lake White Earth supposedly an "historical" Indian name. Interestingly, the Indians also called it Mde Medoza, Loon Lake, which was the name adapted by settlers, or Heyate Mde, Set Back Lake. But, Calhoun *is* an historical name that connects to *our* history. Further, it is unlikely that any Indians in the metro area knew or cared about the new name until a white liberal brought it to their attention. That does not matter because the object is to erase *our* history. The overriding goal of all of this is for every man to become an island, the self-sufficient being whose time has collapsed into nothing leaving him on the same level as an animal—a beast.

This state of being can be seen as a contrast between space and time. Animals are not aware of time; they live only in space. Self-sufficient man orients himself to space, not time. Having no connection with the time of humanity he cares little about history or successors but only about controlling and inclosing his life in a personal space or extending it to control others through war or debt. Having no successors this space oriented man has no future. This is seen most profoundly in the dearth of births in Western Civilization as the time oriented man of Christendom has become a space oriented man of modernism.

Time as Seen by Scientists

Science as it is practiced today assumes a totally determinate universe that started with the big bang and will continue along for more billions of years. A scientist in his lab sets up an apparatus to study a small part of the physical world. He makes a measurement and records it. By the time the datum is entered in his ledger the part of the physical world from which he took the measurement no longer exists. It is in the past. But, he is banking on the fact that the universe is determinate and that the part of the physical universe now present in his apparatus resembles the one he measured in the recent past closely enough to make his observation predictive. That is to say, he is only concerned about time as a numbering of motion. His clocks work and that makes sense of his measurements. There is nothing wrong with that; that is how we lead our lives.

Time as Seen by the Cyclic Creeds

The cyclic creeds such as Buddhism and Hinduism do not include telos in their theology and as a result there is no direction or goal to existance. That is, everything keeps repeating itself. "In the scheme we have set out above [concerning the cyclic creeds] the unfolding of time is a development without substance in which nothing changes because everything changes." [1] This is another case where the refusal to accept authentic divine revelation hampers ones ability to adequately comprehend reality. For reference, Buddhism began in the sixth century BC and Hinduism was a later development of elements of Buddhism and other sources.

World Lines

George Gamow proposed something he called world lines such that each physical thing in the universe has one. By this he means that a thing has a location not only in the common three dimensions of space, but also a location in the forth dimension of time. In that case everyone has a unique world line that goes weaving through this strange space-time continuum. "We see that from the point of view of the four-dimensional space-time geometry the topography and the history of the universe fuse into one harmonious picture, and all we have to consider is a tangled bunch of world lines representing the motion of individual atoms, animals, or stars." [2] For man his world line starts at his conception, and ends when he dies. The interesting thing about this for believers is that everyone's world line ends at exactly the same place, namely their personal judgement before the throne of God. Everyone is given an allotment of time in how long they live and we each must determine what we do with it.

Time Travel

Can there be time travel logically irrespective of the physical possibility? The answer would have to be "no" to travel to the future based of the scriptural references that say the future is known only to the Father. This is also reasonable. The future is determined by the free will of man

---

[1] Henni de Lubar, *Catholicism, Christ and the Common Destiny of Man*, London, Bruns and Oats, 1950, p. 141.

[2] George Gamow, *One, Two, Three. . . Infinity*, New York, Mentor Books, 1947, p. 78.

acting in the present. With all the possible free will decisions that all the people alive at any one time make from instant to instant, the future is far too complicated for anyone but God to know.

As for actual physical time travel, let us start with travel to the past. If one were to do that it would be an event in the lives of the people before now that would change what they did, namely the decisions they made. That would have a ripple effect through all the time from then to the present and into the future. So the present we would have would not be the present we actually do have.

Next, try travel into the future. If it did happen it would mean that the time traveler could not return to the present because if he did it would amount to the same thing as traveling to the past. Since the future is the result of everything that has happened up until the present and all the free will decisions of man and the calamities of nature that will happen to the point in the future where the time traveler landed, he would be one more calamity for the people of that time to deal with. And since the future has not happened yet, and will be determined by countless second to second free will decisions of billions of people, there would be nearly an infinite number of futures into which our traveler could go so the one our time traveler went into would be most unlikely to be the one that actually happens.

Would it be possible to time travel only to look at future times, and not actually be there to partake in the events of the time? The answer to that should be "no" because knowledge of the future, would change decisions made now and the future would be different than it would be without that knowledge. But, does that matter since the future is indeterminate anyway?

Looking at the past we would have to assume that as each *present* disappeared into the past it would stay intact like a frame in a motion picture. That could be, but there is no reason to suppose that is what happens. If that is not the case the past is completely lost to us.

How Long Is the Present?

As we have seen above St. Augustine arrives at the conclusion that the present has no length. This is one of those questions that people rarely think about. Regardless of the length of the present, time appears to flow at a rate that we can for the most part keep up with. After all, *it is our reality*, so why would God make a reality in which we could not function. Think about a speeding bullet. We cannot see it. But a jet fighter plane can fly as fast as a bullet and we can see it. That is first of all because the fighter plane is much larger and further away. And second, there is nothing commonly in nature that is as small and as fast as a bullet that is dangerous to us or that we care about for any other reason. As a result, time moves and our senses perceive at a rate that is compatible with normal life on the average letting aside that time flies when you are having fun and drags when you are in pain.

Getting back to the speeding bullet, we do not see it because the rate of travel of its image across our retina is faster than the image of the jet plane. Apparently there is not time for neurons to fire and send a signal to the brain that something is there in the case of the bullet. Or else, the signals are in fact there, but we do not consciously recognize them. Our concept of time, the only real time, is regulated at a speed that takes into account the physical processes such as the speed of light being very fast, the size of a photon relative to a rod or cone on the retina, and our brain's rate of processing phantasms. Since all non-spiritual concepts come first by way of sense images that cause a phantasm in our physical brains, and since these images are formed at a rate consistent with physical processes, the two are intertwined. That is, there is a harmony between the speed at which physical processes operate and our ability to process sense images. Here we see that the flow of time is keyed to living things in a way that makes sense. If we assume God wanted to make us, and He obviously did, the demands of nature, as mentioned below, would say he had to make the rest of creation to compliment us.

All of this still does nothing to answer the problem of putting a length of time on the present. Clearly the present has something to do with consciousness, and only humans have consciousness. One must put aside sentient consciousness, that of animals, which is another thing entirely. Modern writers like to confuse the two so they do not have to deal with a spiritual soul in man. We know about the present but only through a glass darkly. Time is mostly about the future and the past. Think about making a fist with your hand. First you think about doing it and realize you do know how to do it. You do it and then you have a fist. It happened but then the act is gone, it is in the past. It is gone forever. In fact, if you move your fingers really fast, you hardly see them move in the act of making a fist, especially if your fingers are close to your eyes—the speeding bullet effect.

But, we are still no closer to the length of the present. One thing is certain, it is not long. All of our presents are out there in the future coming toward us with the speed of time. The odd thing is we have some control over those presents coming toward us from the future by how we conduct the present present. If we decide to go shopping this afternoon it will make a different set of presents coming at us from the future than if we stay home. If we knew how many presents there were in a second we would be able to tell how long one was. Maybe they are of all different lengths or maybe they have no length at all as St. Augustine surmised.

Length of the Present as Seen From Theoretical Physics

There is a "constant" that appears in theoretical physics and quantum mechanics that might suggest how short the present could be if it were not zero. That is the Planck time = $(hG/c^5)^{1/2}$ = 1.35 x 10^{-43} second. This time is based on Planck's constant, h = 6.63 x 10^{-34} joule-seconds, the gravitational constant, 6.67 x 10^{-11} m^3 kg^{-1} sec^{-2} and the speed of light, 3 x 10^8 meters/sec. [3] Planck's constant has been considered a building block for quantum mechanics and theoretical physics for a century and the gravitational constant as well as the speed of light have been measured to considerable accuracy. A length of time of 10^{-43} seconds is unimaginatively tiny. Some have suggested that the shortest time physically possible would be set by Planck time and therefore further suggest that the length of the present would not be zero. However, if that were the case we run into the problem discussed by the ancients namely that if it is not zero, it can be divided into part past and part future. As we shall see later, having the present of no extension at all is the better option.

The Starting and Ending of Time

According to modern science the universe was created (they would say *began*) 13.7 billion years ago with the big bang. This is driven by the demands of evolution. But, if it were that long ago or one microsecond before the first rational human mind was created it would not matter. That is because before the first rational mind was created there was no *reason* for time to exist. Animals do not have spiritual souls, therefore are not rational and are not aware of time because they are not aware of themselves or anything. Animals are simply not aware. They could just

---

[3] This information is readily available, cf.: https://www.space.com/what-is-the-planck-time

as well have been created in that instant before the soul of Adam was created and it would not have mattered a bit. There is a saying that takes some time to digest when one is presented with it for the first time. It is this: *Time is a physical quantity that is only used cognitively.*

Of course modern scientists could not allow everything to be created in an instant because creation presumes God exists, an idea they exclude from their science, and further it is not as tidy as assuming a long gestation period for things to get to their present state.

Time Ends with Man

Even if we are not certain that time started with the first rational man, Adam, we do know that it will end with the last one. There are many references to the end things in Sacred Scripture, that is Divine Revelation and they are explicit on the fact that there will be people alive at the end of the present universe. As stated before, this author makes no apology for using Divine Revelation as a source of information since to reject it is short sighted. It is only reasonable to use any and all knowledge available. It is hoped the reader will not be unduly put off for this. Plus, in what follows, it will be shown that one can come to these conclusions with natural reason.

We are told that in the end after a great calamity the present world will pass away and the last judgement occurs. We will see the end of the world and the appearance of the new heaven and new earth and the start of eternity. A few citations follow.

29 And immediately after the tribulation of those days, the sun shall be darkened and the moon shall not give her light and the stars shall fall from heaven and the powers of heaven shall be moved. 30 And then shall appear the sign of the Son of man in heaven. And then shall all tribes of the earth mourn: and they shall see the Son of man coming in the clouds of heaven with much power and majesty. 31 And he shall send his angels with a trumpet and a great voice: and they shall gather together his elect from the four winds, from the farthest parts of the heavens to the utmost bounds of them. 32 And from the fig tree learn a parable: When the branch thereof is now tender and the leaves come forth, you know that summer is nigh. 33 So you also, when you shall see all these things, know ye that it is nigh, even at the doors. 34 Amen I say to you that this generation shall not pass till all these things be done.

35 Heaven and earth shall pass: but my words shall not pass. 36 But of that day and hour no one knoweth: no, not the angels of heaven, but the Father alone. Mat. 24: 29-36. [4]

Looking for and hastening unto the coming of the day of the Lord, by which the heavens being on fire shall be dissolved, and the elements shall melt with the burning heat. But, we look for new heavens and a new earth, according to his promises, wherein dwells justice. 2 Peter 3:12-13.

Whom heaven indeed must receive, until the times of the restitution of all things, which God hath spoken by the mouth of his holy prophets, from the beginning of the world. Acts 3:21.

I saw a new heaven and a new earth. For the first heaven and the first earth was gone: and the sea is now no more. Rev. 21:1.

And God shall wipe away all tears from their eyes: and death shall be no more. Nor mourning, nor crying, nor sorrow shall be any more, for the former things are passed away. And he that sat on the throne, said: Behold, I make all things new. Rev. 21:4-5.

It seems that God the Father has a "time" picked to the day and the hour when the world will end and time will end, and there will still be human beings living when it happens. Therefore, it is wrong to say that at some time in the future due to nuclear war, pestilence, plague, or famine the last human being will die to be survived by ants, cockroaches and rats, and that even after the sun goes supernovae and destroys the earth along with the ants, cockroaches and rats, that the universe will continue on as if on autopilot until it finally runs out of energy and goes away.

Did Time Start with Man?

Thus, having one bracket on time established it is up to us to see if we can establish the other. The easiest way to do that, it seems, is to note that it is only logical that if there is no need for the world (the universe) or time after there are no more people, that there would not have been any need for either before there were any people. This speaks to the the-

---

[4] Mk. 13: 24-32 has almost identical language.

ses that before the first man, a spirit in substantial union with what we call physical matter, there was no need for either the physical universe or time.

We are all familiar with the two creation narratives in Genesis, the first chapter of the Bible. In the first there are six days of creation. This account starts with the heavens and the earth existing and the earth being a formless void. Then God created light, day and night, the sky, dry land and the sea, plants, the heavenly bodies, fish, birds, animals, and only then man. The second account differs in interesting ways. In this one God first created the earth and then the heavens. Then he created Adam followed by vegitation, animals and finally Eve. That is, between the time he created Adam and the time he created Eve he created everything except the raw material, so to speak, the earth and the heavens. Not much room for evolution. That is, God does not allow for time before the first human being in this account. Take your pick of the two but they certainly do not disallow that there is the possibility of no time before humans arrived.

An additional point is that evolutionists use the fossil record as one of the proofs of evolution. The interesting thing is that according to the fossil record homosapiens were the most recent species to appear. That is to say, when man arrived all evolving—if there ever had been any—had already happened. That is, when we arrived, there was the universe and earth with an atmosphere, plants and animals ready and waiting for us sort of like it had just been created. Evolutionists will argue that there must be a species of insect or some such thing some place that evolved into existance since then. But, we have not found it. This leads to the statement that one of the few actual hard facts that we have about evolution is that nobody any place at any time has actually seen it work. The proposition that time started with man will be expanded upon later.

Chapter 11

Human Nature

We must ask the question that if we propose that time is specifically for man or even if man could control the flow of time does he have the power to do so? In this we are, of course, speaking of man's soul. It's important to remind ourselves that a spirit is outside of space and time in its existence. That means that normally it is like oil and water, spirit and matter do not mix, except for us in some mysterious way known only to God. Matter and spirit are simply two different realms of existence. So if our spirit could interact with matter beyond animating our body, there's no reason why it could not affect, control or even destroy matter in other ways, or as far as that goes, control time.

Spirit

We will be speaking quite a lot about spirit so it is well to spend a few words on discussing what, in fact, a spirit is. To start, a soul is the life principle in a living body. Plants, animals and man have souls. Only in man is the soul also a spirit. So what, then, do we mean by spirit? A spirit knows, wills and has power. Our rational, thinking part does not reside in our brains; it is in our spirits. When we say mercy is kinder than justice we are not comparing electrical impulses in our brains. It happens in our souls.

Our ideas are not material. They have no resemblance to our body. Their resemblance is to our spirit. They have no shape, no size, no color, no weight, no space. Neither has spirit whose offspring they are. But no one can call it nothing; for it produces thought, and thought is the most powerful thing in the world—unless love is, which spirit also produces. [1]

A somewhat difficult idea about spirit compared to material things is that a spirit has no parts.

> A part is any element in a being which is not the whole of it, as my chest is a part of my body, or an electron a part of an atom. A spirit has no parts. . . . There in no element of our soul which is not the whole soul. It does a remarkable variety of things—knowing, loving, animating a body—but each one of them is done by the whole soul; it has no parts among which to divide them up. [2]

When we say the spirit has no parts it necessarily means it is indivisible. The reader may think this is a case of belaboring the obvious but it is an important distinction as will become obvious as we proceed.

Since spirit has no parts it does not occupy space since space is what matter spreads its parts into. Having no parts also means a spirit cannot be changed into anything else, nor by any natural process can it be destroyed.

> A spiritual being, therefore, cannot lose its identity. It can experience changes in its relation to other beings—e.g. it can gain new knowledge or lose knowledge that it has; it can transfer its love from this object to that; it can develop its power over matter; its own body can cease to respond to is animating power and death follows for the body—but with all these changes it remains itself, conscious of itself, permanent. [3]

That spirit has power beyond what we find in the material universe becomes apparent only when we look closely at the result of a being that is partless. Having no parts means first that it has nothing in common with the material world. It also cannot lose any part of itself the way friable matter can. But more importantly, it is ontologically superior to and more powerful than matter—all matter—due to the total integrity of its being. It takes a fair amount of contemplation for that to finally make sense.

---

[1] Ibid., Sheed, p. 8.
[2] Ibid., pp. 8-9.
[3] Ibid., p. 10.

In our effort to make the concept of spirit wholly our own we must be very leisurely. We must keep on looking at the relation between having parts and occupying space, until we find ourselves quite effortlessly seeing that a partless being is outside space. We must keep on looking at the meaning of being and the meaning of part until, again quite effortlessly, we see that such a being must by its concentration be superior in essence and in operation—that is, in what it is and what it can do—to those others whose being and operation are dispersed. [4]

The Power of the Human Spirit

In the book *Occult Phenomena*, the author, Alosi Wiesinger, writes about the powers of the spirit we posses as our souls. The reader should understand that in *Occult Phenomena* he looks at "occult" from its real meaning which is "hidden," as when the moon occults the sun shining on the earth, that is, it hides it, we have an eclipse of the sun. This is opposed to a common meaning where occult is associated with the fallen angels. As such, the book deals with spiritual phenomena that are normally hidden and that are only occasionally encountered.

The faculty of sight, for instance, is in the eyes, but the soul's capacity for cognition is not confined to any one part of the body; indeed in this respect the soul is not only not wholly present in every part of the body, but not wholly present in the body as a whole, for the power of the soul exceeds in its activity the capacity of the body. When therefore I speak of the partly body-free soul, I am not suggesting that there is a substantial separation from the body, but that its purely spiritual powers reach beyond the body's domain and that in this way it is empowered to perform feats in which the body has no part, or simply an abnormal one. [5]

As we see animating our physical bodies is a simple task for the power of a spirit, any spirit including our souls. One of the powers of the soul that each of us has, but rarely think about, is our subconscious. He

---

[4] Frank Sheed, *God and the Human Condition*, New York, Sheed and Ward, Inc., 1966, p. 74.

[5] Alosi Wiesinger, O.C.S.O., *Occult Phenomena*, Westminster, Maryland, The Newman Press, 1957, p. 56.

suggests that our subconscious holds far more memory than our conscious memory does.

> Everything that flows towards the soul from the outside first enters into the subconscious, and from here only a small part goes into the upper consciousness at all. The subconscious is therefore much the richer of the two; it leads an independent life, being, so to speak, "busy behind the scenes". It can thus provide an explanation for much that seems to us incomprehensible and surprising. Though everything does not penetrate into the upper consciousness, yet nothing is lost. Experiences may only enter the consciousness after delay, or even not enter it at all, yet they remain effective and condition the freedom of our actions—or they have the effect on us of an alien intelligence. [6]

He also suggests the other occult phenomenon such as psychokinesis—the moving of physical objects with the mind—is simply a manifestation of the power of our sprits. His premise is that when these paranormal powers are exhibited, they are the peeking out of some to the preternatural gifts that were lost to humans by Original Sin.

According to Wiesinger, prior to the fall, Adam had the powers of a pure spirit.

> By his angelic intelligence Adam knew how to avoid the causes of death and disease and by his will he was able to direct the fluid and solid substances of this world, so that they not only did him no hurt but greatly contributed to his happiness. [7]

He expands on this by saying that our first parents would have had complete power over material substances.

> Our first parents possessed the preternatural gift of a spiritual will which reached out beyond the body, a will which gave man the power of acting on matter and moving it without any kind of effort, even as pure spirits can act upon it and move it. We may thus suppose that Adam performed bodily work for so long as this gave him pleasure and redounded to his health. When, however, it

---

[6] Ibid., p. 61.
[7] Ibid., p. 84.

threatened to become wearisome, he used his angelic powers over matter, as he required them. [8]

It seems reasonable that if Adam could control matter, that is move, mold and in some ways change it the same as angels can, he would have had the power to control time as well. Now, angles are not subject to time. "Angels . . . [have] endless duration without the vicissitudes of time." [9] So angels would not have a reason to control time. But, Adam did live in time and if he could control everything else, why not time?

So, in answer to our question, it seems reasonable that before the fall, Adam and Eve would have been able to control the flow of time if there had been any reason to control it. Adam could control matter to avoid diseases and to help with his work when he became weary. There seems no reason for him to control time, though. And even if Adam could control time before the fall, can we collectively as the human race, do it now after the fall?

Triune Matter

If all of this seems a little hard to believe, we might as well make it worse by considering one of those unique things about matter that is necessary for us to exist. It is well proven in experimental physics that subatomic particles can have both wave and particle properties. That is, in some experiments they behave like they were particles—matter—and in others like waves—energy. In a similar fashion electromagnetic waves, such as visible light, also appear as waves in some cases and particles in others. This defines a duality that could be said to mean that in some cases these entities appear as matter and in others as energy. Further, as we have seen in atomic power plants and nuclear weapons matter can be converted to energy and vice versa and there is an equivalence expressed by Einstein's famous equation $E = mc^2$. Beyond those two manifestations of matter there is a third.

The quality of spiritual impulses, ideas, plans, will-power, attitudes, etc., cannot be measured with physical-chemical methods, (even though these ideas, etc. use physical-chemical carriers of information), because these spiritual qualities are more than merely

---

[8] Ibid., p. 85.
[9] Ibid., Glenn, p. 264.

physical-chemical. Reducing reality to physical-chemical appearances therefore, is an arbitrary act of faith which is unusually unscientific. . . .

In spite of the fact that thoughts and impulses of the will cannot be measured it is apparent that the elementary particles—atoms, molecules, cells, and organs of our bodies—respond to the will and spirit, and carry out their wishes. It is also apparent that a [magnetic] tape is not changed quantitatively when it receives a recording, but is electromagnetically changed qualitatively so that it can transmit a new quality. The qualitative power of our elementary particles is possible only when they possess a third form of appearance. A qualitative form, appropriate to the spiritual form, which makes responses to the spirit possible, must exist over and above the electromagnetic quality of a wave or a particle. This is a quality to which the waves or particles can respond. The physicist, Bernard Philberth, explained this type of understanding of living matter, with its three forms of appearances as a "triune," "trinitarian" dimension, in his book *Des Dreieine*. (Bernard Philberth, *Des Dreieine* Christiana Verlag, Stein a. Rhein, 1974.)

We experience the power of our physical-chemical energies to respond to spirit and will in the activity of our brains. If these energies are capable of reacting to spirit in the brain, they are, in principle, also capable of doing so in other parts of our organism. [10]

The existance of triune matter is seen in practical terms when we, for example, have a thought in our minds that we want to move our hand and then do it. There has to be a connection made from our spirit to the matter or energy in our brains to produce the chain of physical-chemical events that cause the muscles in our arm to contract as directed. This is psychokinesis—controlling or moving matter with our minds—on a micro level. And, we all do it all the time.

This means that we all have the power to move or affect at least a little matter with our minds. As we discussed earlier, the animation of or physical bodies is a ridiculously simple thing for the power of a spirit. Therefore it is not unreasonable that the collective human race plays a substantial role in keeping the physical universe in existance.

---

[10] Siegfried Ernst, MD, *MAN The Greatest of Miracles*, Collegeville, MN, The Liturgical Press, 1976, pp. 148-9.

God Looks at the Human Race as One Thing

If we do control the flow of time now, would it be done by us collectively, that is all of us alive today, and all who have lived from Adam to those alive at the end of time? One way to answer this is to note that in some mysterious ways God looks at the human race as one thing. This is especially seen in the mysterious idea of original sin. Adam sinned and we all inherit that sin, or one might say that condition, from him.

> Obviously there is something in the solidarity of the whole human race clear to God but not to us, that He could so treat the race as one thing. . . . God sees the whole race, every member of which He created, as one thing—somewhat as we see a family as one thing or even a man. The mere number and variety—myriads upon myriads of men—and the uncountable ages, do not impede the vision of the Eternal and Omniscient God. [11]

There is another way that God looks at the human race as one thing. That is in the general judgement at the end of the world. Why have a General Judgement if each one of us is judged as to going to heaven or hell at his personal judgement when he dies? It is to show the absolute justice of God's judgements. We will see everything everyone who ever lived did and how it affected everyone else, especially us. That is, we all affect everyone else. "The Last Judgement will reveal even to its furthest consequences the good each person has done or failed to do in his earthly life." [12]

Concerning the human race as one thing from the perspective of original sin a quote from Wiesinger's *Occult Phenomena* will follow.

> Original sin, the sin of our first parents, inherited by all their posterity, consists formally in the deprivation of sanctifying grace with which man had been endowed by God and which he lost both for himself and for the whole human race—as indeed is plainly stated in St. Paul (Rom. 5. 12): "As by one man sin entered into this world and by sin death . . . so death passed upon all men in whom all have sinned." . . .

---

[11] Frank J. Sheed, *Theology for Beginners*, New York, Sheed & Ward, 1957, p. 57.
[12] Catechism of the Catholic Church, 1994, § 1039.

The difficulty, as it seems to me, is not that all men should be punished. . . in the case of original sin, however, we are not only all punished, but we are all guilty. . . .

The difficulty becomes even greater when the theologians tell us—and quite rightly—that original sin must be for us a free act of the will. . . . It must be a free act of the will if it is to be a real sin at all, even if it is only an habitual state of fallen nature, because sin is a free and knowing transgression of a divine command. How then can it be that original sin is a free act of the will for us? . . .

St. Thomas (De Malo, q. 4, a. i) says that man must not be treated as a single person but as a member of the human race which has its starting point in Adam, as though all men were a single man. . . . How can they psychologically represent one will in such a way that original sin would become a free act of every member of the race?

The only way of giving a certain answer to this question is to refer back to the pure spirituality of our first parents, a spirituality which would in part have been inherited by their descendants; to the latter there would also have passed that capacity for being influenced, that noopneustia, of which the writer spoke when he showed how angels partake of the knowledge of angels higher than themselves by illumination, and having partaken of that knowledge, obey them. They are influenced with a degree of power which we simply cannot imagine. . . . This noopneustic power rested in Adam who would have been spiritually one with his son (who in his turn would have been similarly one with his own children) and would so have influenced that son that he would have been wholly obedient to his father's will. This will would have been passed on from generation to generation, and would have determined the wills of posterity precisely as the wills of the higher angels determine those of the lower ones—or as the will of the hypnotist influences the will of his subject. Thus we would have been born with the same disposition of will as Adam possessed. . . . Any deviation from this, though physically possible, would have been impossible morally, or would at the most have only been possible in matters of little importance, in so far as this was necessary for the assertion of free will. This accord would have been firm, instantaneous and irrevocable, of the kind we have already noted in the case of pure spirits. Thus the will of posterity was actually contained within the will of Adam, so that his sin became our own, Adam's posterity was infected, "being prone to evil from . . . youth" (Gen. 8. 21) and "sold under sin" (Rom. 7. 14). Adam's sinful act thus became actually morally and psychologically our own. . . .

The dividing wall of individualism was necessarily a consequence of sin. In this way the Catholic doctrine of original sin provides an indication that our first parents, in addition to their human nature, also possessed as the basis of their preternatural gifts that of pure spirits together with all the faculties appertaining to the latter which we have enumerated above. [13]

As mentioned, the original sin affected everyone who will ever live. We also see certain historical figures who have affected history and hence the lives of all who come after them. Examples would be Julius Caesar, George Washington or Adolf Hitler. In one way or another their actions changed the course of societies and hence all who came after them for good or ill.

Just because the things we do in our seemingly little lives are not known as historical events does not mean we do not affect others in profound ways. And, these effects have a ripple effect. The concept of six degrees of separation comes to mind. That says that if you take all the people you know, and all the people each of them knows, and each of them knows and do that six times you will include all the people in the world. This has been tested in a reduced way with the Internet and found to be more or less true.

The point is we are all sort of one, like the fibers in a vast cloth. The fibers at either side of us could be compared to the width of the fabric and are those alive today. Those closest to us we touch. In turn they touch those next to them and so on. But, we are also connected to those behind and before us. That bolt of cloth is the human race. And that single thing is what controls the flow of time. Years ago we had an old oak tree in our back yard that was quite rotten inside. My boys joked that during a strong wind it was only the carpenter ants inside holding hands that kept it from falling down. The human race may be something like that.

---

[13] Ibid., Wiesinger, pp. 85-89.

The Demands of Nature

If man has the ability, that is the strength, to control the flow of time, and we do it collectively so it is a smooth continuum, is there any reason we do control time rather than letting God do it as part of the physical world?

That man needs time is rather self apparent. There are what philosophers call "the demands of nature" and time would fit into this category.

What is it that we understand by nature and the supernatural? We call all that "natural" which constitutes a substance, or derives from it or which demands it. This means:

1. All that inwardly constitutes the specific essence of a thing, whether it be an essential or an integrating part of its being.

2. Everything that proceeds spontaneously from the nature of a thing, such as aptitudes, talents and powers, and everything that can proceed from it under the influence of some other being, such as proficiency in some art, skill or craft.

3. Everything which, while lying outside the thing itself, is nevertheless necessary for its continued existence (nourishment, light, air), for its activity (the God-given will for survival), for its development (instruction, society, state) and for the attainment of its goal (knowledge of God, free will). The theologians group all these together under the term "demands of nature" or of **things due, the things that God had to allow men to have, assuming that he desired to create men at all**. Emphasis added. [14]

One could certainly include time in category 3 above. In this case, time would be "necessary for its [the human race's] continued existence," but it says nothing about the control of it any more than we control light or air.

The question is, can time be included in either category 1 or 2? If it can, than the control of time would seem to be one of the "demands of nature" for man. Let's look at category 1 first. It seems time constitutes an "essential or integrating part of its being" for man. But, this says nothing about the control of time. Looking to category 2, is the control of time one of those things that "proceeds spontaneously from the nature of a thing [that is, us] such as "aptitudes, talents and powers. . . ?" If it is, it

---

[14] Ibid., Wiesinger, p 75.

would come under the category of powers rather than aptitudes or talents to say nothing of an art, skill or craft. If we have the power to control time, this definition is silent about it.

The whole point here is God created us in the odd way that he did. We need time, so he created time as well as the rest of the physical world. So as not to slip into an existential frame of mind, there is a real world out there and there is real time; that is, it exists independently of us.

However, there are small parts of that physical world out there which we can control as when we build a house with a fireplace to make it warm inside when it is cold outside. We do not make the energy that is produced in the fireplace by the burning wood and contained around us by the walls and roof of the house. Yet, even though we do not make the energy, we do exert control over the flow of some of it to keep us warm. The problem with time is we cannot control the flow of a small part of it. It seems to be all or nothing. As far as we know, time is a single thing for everyone. It is not a matter that some people have a lot of it and others very little at a particular instant, i.e. the present. This is not to be confused with the "length of time allotted to each individual" analogous to a piece of string: some people have a much longer piece than others. The point being that the string is the same thickness for all. And, the flow of time is the same for all.

If there were only one human he could be expected to cause the flow of time to be rather erratic, as we sense such a great variance of the flow of time as in we have no sense of it at all when in a deep sleep. Other than that, it is not likely that the single person, if there were only one left, could measure the erratic flow of time. But, with the collective sense of the flow of time of billions of people the average is consistent.

From the above it seems we can say that being able to control the flow of time is not a "demand of nature" unless . . . .

Controlling the flow of time by people would not be a demand of nature unless nature could not be counted on to do a proper job of it. After all, we have established that time exists *for* us. Now look at the rest of nature that was created for us. But, there are storms, drought, pestilence, earthquakes and all manner of untoward happenings in the physical world that are not beneficial to man. In spite of this, we are still allowed a measure of nutrition, air, light, etc. or there would be no people. But, the way God has set it up, each of us gets these physical needs in different measure—some get sick, some do not, some are well fed, others starve. Using our free will we make an effort to smooth out these differences by planting crops, building houses and doing all the other things an organized society does. But time, as far as it applies to those living at any one instant, appears to affect all the same. So, is time one part of the physical world where God has made an exception—does He let us control it collectively? It could be it is in the control of time that we find the one place where the entire human race is still in harmony after the sin of Adam. It makes a sort of sense that we need something left to us that we all hold in common.

In this line of reasoning we have examples from society. The United States was made up from the beginning of people from around the world. We were called the melting pot. That is, many peoples came here but they all eventually became Americans. Now, we are becoming an extremely diverse society where the various minorities are not melting into American society. Liberals will say that our strength is in our diversity. That is a fundamentally wrong idea. No society can exist unless all members hold at least some things in common. In the past one of our common things was language—English. Another was the Christian religion. Those are now being lost as a common denominator and we are losing our identity, our cohesiveness and our freedom.

To continue. On the other hand, one could object that we all do, in fact, get a different measure of time in that we all have different life spans as mentioned above. That could in turn be answered by saying that the other physical needs are not measured by the total amount, say food, one gets in a life time, short or long, but in how much is allocated *as needed*. So, from that point of view we should each get different amounts of time *as needed*. It appears that nature does not do a proper job of controlling the physical world, and we, using our free wills, can control parts of it, so why could not we, using our free wills, control at least parts of time?

Chapter 12

Man Controls the Flow of Time

There is relatively little mentioned about what controls the flow of time. We have mentioned that it seems to flow at a rate that is consistent with the rate of the sensory perceptions of human beings. It is also obvious that subjectively it flows at different rates as in periods of happiness it seems to fly along while during periods of loneliness, sadness or pain it seems to drag. But, for all people taken collectively, it appears to have a steady rate of flow. We will leave out of this discussion the relativistic effects on time postulated by Einstein.

The control of the flow of time is considered here because if it can be shown that people control the flow of time it would go a long way to showing that time only exists for us.

To prove the conjecture that people control the flow of time requires the establishment of three premises. The first is that time only has meaning for human beings and as a result there was no time before the first human, commonly known as Adam, was created, and that there will be no time after the last human dies. That is, time was created, or if one prefers exists, for us. This has already been covered to some extent. The second premise to show is if we want to prove that man controls the flow of time is that he has the power to do so. The third, then, is to show that if time were created for us, and if we have the power to control its flow, that we indeed do collectively control that flow. This proposition is opposed to what everyone has always assumed which is that the control of the flow of time is something God does along with making the sky blue on a clear day. Or to put it another way, is the *control* of time—not just the existance of time—a "demand of nature." That subject will be discussed along with others as we proceed.

We begin by asking the question of whether there is some force or power that regulates the flow of time. Gardeil refers in passing to the fact that there must be such a thing.

> This much, however, may be said. In the matter of time, as in the theory of place, the Aristotelian position presents, no doubt, certain aspects that will not stand up in the light of contemporary scientific thought. But other aspects have proved more durable. The idea that cosmic motion is a unitary system, *or that a regulating principle of time is necessary*—these, if not others, are far from gone. Emphasis added. [1]

Here we will consider the proposition that it is man, the collective human race that is living at any one time, that controls the flow of time. But first it would be good to explore a few simple philosophical principles that would be needed to foster such a claim.

Causes and Agents

In the preceding chapters some concepts were presented with the primarily intent of getting the reader predisposed to think about time. In the following pages those ideas along with others will be presented in the hope of reaching some useful conclusions.

It has been shown above that time is a demand of nature for man. The question is open as to whether man controls the flow of time. Here we will look at time as a cause or an agent to determine whether or not this will shed light on the control of time or even on what time is.

From various places we get the definition of agent as an entity that is capable of action; a person or thing that performs an action or brings about a certain result, or is able to do so; an active force or substance producing an effect as a chemical agent.

As for causes, here are four main types of causes recognized by philosophy: material, formal, efficient, and final. Take a marble statue as an example. The block of marble is the material cause, that is the material out of which the statue will be made. The finished statue is the formal cause as it gives it its from. The sculptor is the agent that produces the statue and is called the efficient cause because he produces the form. In the case of the sculptor the final cause is the reason the sculptor made the

---

[1] Ibid. Gardeil, Vol. II, Cosmology, p. 126.

statue as for example, fame, money, or simply to make something beautiful. In the more general case final cause explains the direction of things
to an end. This is called teleology. In only one of the causes do we meet
an agent and that is as an efficient cause. The sculptor operates outside
the material and the form. To say it another way an *agent* is the capacity
to *do* and a *cause* is the capacity to *be*. Yet in normal parlance we think
of a cause only as an agent as in, "He caused the auto accident."

Agent Causes

Now, we must investigate agent causes. To begin, we must distinguish between movers and agents. A mover changes a property or accident of an object as in changing its position or velocity. That is, physical
changes are correlated with movers and changes in substance is correlated to agents. As a result when above we identified efficient cause we
were in other words saying an agent cause. Frequently movers and agents
are used interchangeably but as seen here there is, if we use precise
terms, a difference.

"Thus the form or internal substantial structure comes to be in the
potentiality of the fundamental material and depends upon the disposition of properties brought about in the antecedent substances upon which
the principal agent acts." [2] In the case of the statue the block of marble
was the antecedent substance and the principal agent was the sculptor who
brought about a new substance by producing new properties in the marble.

Self-Evident Propositions

There are several principles of reason that we as humans use in rational thinking. These are called self-evident propositions or sometimes
analytic propositions. They cannot be proven, only explained. They are
of the nature that if you understand the terms used in the proposition you
understand the proposition. For example: The whole is greater than any
of its parts. If you understand the words, you understand the truth of the
proposition. A short list of self-evident propositions is as follows.

- Principle of contradiction: Anything either exists or does not exist
  (a thing cannot both be and not be in the same way at the same
  time).

---

[2] Richard J. Connell, *Nature's Causes*, New York, Peter Lang Publishing, 1995,
p. 140.

This one has the following corollaries.
  • Principle of the excluded middle: Anything either *is* or it *is not*.
  • Principle of identity: Whatever is, *is*, and that which is not, *is not.*
  • Principle of difference: *That which is* is not *that which is not.*
- Principle of efficient causality: That which is not from itself is produced by another.
- Principle of finality: Every agent acts on account of an end.
- Principle of reason for being: Whatever is has an explanation or sufficient reason of its being.
- Every agent acts for the sake of an end.
  This one also has corollaries.
    • Those things which are unto an end are not good save in relation to that end. For example, a carburetor on a car engine is only good when it is used for its intended purpose. It would not make a good stapler nor a good flashlight.
    • The imperfect is for the sake of the perfect.
    • The lower is for the sake of the higher.
    • The part is for the sake of the whole. That is, the carburetor is made for the sake of the engine.
- The whole is greater than any of its parts.
- The greater does not come from the lesser.
- Two things equal to a third are equal to each other.
- No cause can produce an effect beyond its—the cause's—actual capacity.
- Anything received in any way whatsoever is received according to the limitations of that which is received.
- Good is to be sought and evil avoided.
- Two solid objects cannot occupy the same space at the same time.

This last one mentioned is not always included in the list but is a good place to start. When you are out Christmas shopping and there are no parking spaces around the shopping center that one does concern you. In frustration you drive around saying, "I can't find a parking space." That is not philosophically correct. You can find many parking spaces. What you cannot find is one not already occupied by another solid object. And since your car is also a solid object you know that trying to park in one of the spaces already occupied by another solid object will not lessen your frustration. Anyone will immediately accept this self-evident proposition as true. But when you say, concerning evolution, that the greater does not come from the lesser evolutionists will out of hand reject it.

The reader may ask why he has been subjected to self-evident propositions, something he was not even slightly in the mood for. The reason is this: rather than simply pop one of them out of nowhere this writer thought it would be good to give a short general description of the subject so as not to seem to be inventing an idea to suit his needs. That being the case the self-evident proposition that concerns us here is the Principle of finality: Every agent acts for the sake of an end.

Man Controls the Flow of Time: The Instrumental Agent Case

We could say that time is an agent that operates outside of man, and acts for the sake of an end which is man. It is not an efficient cause—that is agent—as in the case of the sculptor; it is an instrumental agent as in the chisel that the sculptor uses. Now, it would be wrong to say, continuing with the example used earlier, that the engine determines exactly how the carburetor must be designed, though what the carburetor must *do* for the engine will set limits on how it *can* be designed. In similar fashion, what time must do for man sets limits on how time operates just as, from our other example, the sculptor sets limits on how his chisels are designed, and how they are operated, that is, how they are used to form a statue. The chisel is an instrumental agent only insofar as it is moved by a principal agent, the sculptor. Since time is only for man where it is an instrumental agent for him, he becomes the principal agent (efficient cause) moving it and hence controls the rate at which it flows.

The paragraph immediately above is a bit compact so some expansion of the idea of time as an instrumental agent may help make the concept a bit more natural. We quote from Connell the following:

> An artist who sculpts a statue out of wood obviously must employ chisels and mallets and perhaps other instruments without which the statue could not be produced. Each of these has its own design and function, and to the extent that the artist uses them for removing chips or slices of wood, all he does is apply the tool to its own function. But in addition to its application, each tool receives from the hands of the sculptor a modification of its motion through which it is directed to the production of the effect that is more than a separation of a piece of wood from the main block. Clearly the

production of a likeness of Socrates in wood or stone presupposes something more than the chisels and mallets with their inherent design; and so we assign the image-effect to the sculptor because only he is proportioned to the effect that comes about. Over and beyond their design and application, the instruments receive special, directed movement through which they produce the likeness of Socrates, the ultimate effect.. . .

The artist first conceives the likeness of Socrates in his mind and imagination and then executes it. The artist is proportioned to the effect and is able to bring about both the effect and the means though intellectual conceptions that direct the productive actions, whereas the instruments by themselves are not so proportioned. And so to repeat: over and beyond the application of their functions, the instrument must receive from the sculptor special motions by which the likeness is produced. Moreover, the special motions are transient: they are motions that do not permanently modify the design or character of the instrument-recipients, existing in the instruments only while the artist is acting upon them. Instruments, then, are agents only while they are moved by a principal agent, and *only when they bring about an effect that is beyond their own character or nature, their own operational abilities*. Emphasis in the original. [3]

In the case of time, when it is used as an instrumental agent by humans it brings about an effect that is beyond its own character or nature which is awareness in man. This is so because with no time there would be nothing for man to be aware of since the universe would happen in a timeless instant.

It could be argued that in the above example after the artist finishes making the statue and lays down his tools the tools still exist. Since time exists only for man and will go away when he does, it makes time a special kind of instrumental agent. As such, "the special motion (rate of flow of time in this case) is transient: its motion does not permanently modify the design or character of the instrument-recipients (time), but exist in the instrument (time) only while the artist (man) is acting upon it." In this special case the instrument goes away when the artist (the human race) stops acting on it.

---

[3] Ibid., pp. 156-7.

In all of this there are three things at work: the artist, the tool (instrumental agent) and the work (statue) to be produced. In our case we have the artist (man), the instrumental agent (time) and the end work which is each particular life of living humans which all have awareness as an essential quality. To repeat our initial assertion based upon a self evident proposition: time is an agent that acts for the sake of an end which is that strange aware thing that exists in space and time called man.

For an analogous example of this where the instrumental agent goes away when the "artist" stops using it, consider the following. A plumber needs to solder copper pipes together to form a water tight joint. He uses a propane torch, or more exactly a propane flame. The heat from the flame causes the solder to flow and seal the pipe joint. In this case the instrumental agent, the flame, does not exist until the plumber lights the propane gas, and it goes away after the valve on the torch is closed. The flame is also an agent that does no good if the torch with lit flame is left setting on the workbench. What would be the point! It is only an instrumental agent when the plumber directs the flame to the pipe joint. So, it is reasonable to say that time is an instrumental agent that not only receives its special, and transient, motions from the principle source (man) and that that instrumental agent ceases to exist when the principle agent ceases to use it, i.e. when the human race in this physical world is ended.

Man Controls the Flow of Time:
The Awareness Case

The issue of whether people control the flow of time must be answered because having it flow much slower or faster then it does would not work for us. In what follows the idea of awareness is key so some words should be said about it. To begin with, we do not mean the common use of the word as in someone asking, "Are you aware of the fact that the government is not honest when it publishes the official amount of inflation each year?" Anyone who is not aware of the fact, that is know, that the government lies is probably not human so will not be reading these words.

What we mean here by awareness is a fundamental property of humanity. It is the ability we have to say, "I exist," or "I am this person, and not just any person, but this special person." This awareness is identified closely with if not identical with ego. Animals do not share this capacity with us. By way of illustration there is a large elm tree in the neighbor's yard. It is home to a pair of gray squirrels. Frequently one sees them chasing around the tree as if playing a game of tag. However, that is no game. They are defending their nesting territory. As a somewhat comical example of this one day an albino squirrel that was totally white was in that tree. There was a mad scramble all through the branches as the two chased the one. The stark difference on color made it easy to keep track of the antagonists. The normally colored animals seemed to be saying, "Hey dude, this ain't no integrated neighborhood. Get you white tail out of here." But, color had no bearing on what the pair of squirrels were doing.

The reason for the example is to illustrate that neither awareness nor ego has anything to do with the activities of squirrels. It was simply an instinctive reaction to intruders which has survival advantages. A heavy concentration of squirrels would unduly draw predators and just as importantly the pair of squirrels did not want the local food supply to be depleted by interlopers so they would have enough nourishment to bring their offspring to maturity.

However, one would expect the evolutionist to say that awareness has nothing to do with evolution and that time was here before people arrived. Their contention would be that on the large scale there are two main controlling time periods which are the day and the year and both are determined by the scale of astronomical bodies. On the subatomic level the various particle motions like the speed of electrons around the nucleus of an atom is another fundamental motion and it too seems to be controlled by the scale of things. Therefore, as life evolved it did so in tune with these natural rhythms. Many plants and animals have a reproductive cycle keyed to the year. That is to say, people, as animals, have their life processes set to the speed of time that was already existing. It is not surprising that they would make that argument.

It is worthy of note that the day and year as experienced on earth are unique to the earth. At the very least it would be an argument that there is not likely to be life elsewhere in the universe because our day and year combination would be exceedingly rare. There are books written that describe dozens of unique things about the earth and missing any one of them would make life impossible. This is called the anthropic principle. Therefore, let's say the earth is unique. That in itself does not mean that there could not have been non-rational life after the unique motions of the earth were established. We must go to something more fundamental.

It comes down to the idea that before the first human there was no reason for anything else to exist. God does not do frivolous things. There would be no reason in His creating anything that did not have free will, that could say yes to Him or no to Him. What purpose would be served by having a universe full of matter and energy that had no choice but to do what God built into them to do. If one knew all the stresses in the earth and the physical laws that controlled them every earthquake could be predicted to the last second. If one knew the precise instincts that controlled the behavior of an animal as well as all of the physical and biological laws at work, the actions of any animal would be known from moment to moment. This results in an entirely determinant universe from start to end unless man with his free will is interjected into it. Why would God make a thing for the sake of having a thing?

God could watch the universe go from the big bang to the final fizzle as all energy and matter ran down to a final lump of entropy. If it took thirteen billion years—this is the accepted time used by evolutionists and many physicists—to get to where we are now, it will take, say, another hundred billion years to reach that final state. But, for God there is no time so it all happens in a timeless instant. What's the point! So, we say again, time is pointless without a free will being.

Therefore time flows at a rate that makes sense to the way God created the first man who was the first material creature that was aware of time because he was the first material creature that was aware at all. *It is man's awareness that makes time anything but a timeless instant.* That means that there was no time and hence no universe before the first rational animal with an "aware" mind existed whenever that happened. If it happened at the time of the Big Bang then the thirteen billion years really did happen, though, it would seem we should have "evolved" a lot more than we have in all that time. In any case that is nonsense because as mentioned before, the demands of nature for man could not be met in outer space nor in the center of a star. That means that time certainly did not start before a planet identical to ours existed with an environment in all respects similar to the one in which we live and that place contained at least one aware creature.

Time flows for man and not for rocks, trees or animals. So, does the collective human race control the flow of time? Yes, because it is for them and why not have it flow at a comfortable rate for them. It is similar to someone being the only occupant in a room and this room having a thermostat specific to it. Why wouldn't that person set the temperature of the room to that which best suited him? In the case of where there were fifty people in the room one could ask each one what temperature he preferred. With that information a median temperature could be established and used as the most suitable. Notice median would have to be used so half of the people would want the temperature warmer and half cooler than they would prefer. Using an average would be unadvisable because one individual might have suicidal tendencies and say he wanted the temperature at ten-thousand degrees.

In the case of time once again one could at first assume there is only one person in the universe. If this were a very energetic person who was always bemoaning the fact that he could not get enough done in a day, and if he could control time, he might slow it down so as to accomplish more. Or if the person were lazy and lamented the long boring day, he might speed up time so his boredom would pass more quickly. But, since there are many people, each pulling in his own way a rate for the flow of time is reached that is usable by everyone.

If this argument is not fully satisfying to the reader, it must be borne in mind that the awareness of man *is* the present. That plays into what was said above and will also be key to what follows.

Man Controls the Flow of Time: The Act and Potency Case

We have now considered two ways that man could be considered as having control over the flow of time—the instrumental agent case, and the awareness case. These have their merits but may leave the reader still uncertain. Here we will consider the proposition from a more fundamental philosophical point of view.

We find the concept of act and potency is one of the most basic, if not the most basic, ideas of metaphysics. They are considered a single idea because it is difficult to think of one without the other. Potency belongs to those primary analogous notions that cannot be defined in the proper sense of the word. As such, one must proceed inductively, by examples of it, as well as by comparing it to its opposite—act. Here will be given treatments of these concepts by contemporary authors.

Actual Being—Potential Being.—Here we have determinants of real being. A real being that exists is *actual* being. A real being that can exist but does not, is *potential* being. In so far as anything exits, it is actual; hence, actuality is a perfection. Insofar as anything existible does not exist, it is potential; hence, potentiality is an imperfection; it is unfulfillment. This is way Aristotle defines

God, the Infinite Being, as *Pure Actuality*. The transit from potentiality to actuality is called *becoming* or *motion* or *change*. There are four chief types of change; change of substance or *substantial change* (as from living body to dead body; as from lifeless food to living flesh); change of *quantity* (as growth or diminution); change in *quality* (as from hot to cold, from ignorant to learned); change of *place* or local change or local movement. In point of change we see illustrated the axiomatic truth that nothing *becomes*, nothing passes from potential to actual, except under the influence of what is already actual. Emphasis Glenn's. [4]

Notice that the statement in the paragraph above, "The transit from potentiality to actuality is called *becoming* or *motion* or *change*" is nearly the same as the definition of time, ". . . motion or change under the aspect of before and after, is the basis of *real time*." [5]

Another discussion of potency and act is as follows:

The equivalent thought emerges from another Scholastic axiom, that what is in potency cannot be raised to act unless by a being in act. Potency, this means, cannot be raised to the level of act by itself; only a being already in act, exercising efficient causality [efficient cause, one of the four causes discussed above] can bring it about. Yet, though necessarily in act, the agent must also have the potency to act. Is not this a contradiction? No, it is quite possible to be in act and potency at the same time, but not in the same respect. In a created agent act and potency are complementary. For the exercise of efficient causality the agent must be in act through possession of the form (or perfection) which is to be produced in another; and must at the same time be in (active) potency as regards the action to be performed. So, for instance, the intellect when actuated (or in act) by the impressed species, is in (active) potency as regards the act of intellection. [6]

That is to say, when the intellect is actuated by the impressed species of the image of a tree, it is in potency as regards the act of forming the notion of treeness.

---

[4] Ibid., Glenn p. 234-5.
[5] Ibid., p. 264.
[6] Ibid., Gardeil, Vol. IV, Metaphysics, p.194-5.

Both excerpts contain the fundamental axiom: *potency cannot be raised to the level of act by itself, but only a being already in act, exercising efficient causality, can bring it about.* Act must come before potency or no potency would ever manage by itself to proceed to act. It also means that the being in act must be proportioned to the effect, that is, the act, it wishes to bring from potency. These statements will have a direct bearing on the case to be made for time as we shall see.

To better understand act and potency several short quotations will be presented.

The co-principles of matter and from are distinguished as act (form) and potency (matter). [7]

The statue did not exist in act in the naked marble, but it could be hewn because it was there in potency. In the fabrication it went from a statue in potency to a statue in act. . . . Change, every change, we shall find, is a going from being in potency to being in act. [8]

. . . now, whatever the instance, we find one thing common to the state of potency, namely relation to act. Potency always expresses relation to act. [9]

Act is prior to potency, hence not necessarily attended by a potency. [10]

Thus, in the composition of act and potency act is related to potency as the limited to the limiting. . . . Consequently perfection, is limited by a principle distinct from yet united with it; this principle is potency. Thus, in every composition of act and potency, act is limited by potency. [11]

Potency denotes capacity for perfection, whereas act of its very nature signifies perfection; such opposite notions must correspond to entities that are really distinct. [12]

Act and potency are not two beings but principles of being which determine each other but which while really distinct, do not have each a distinct existance. [13]

---

[7] Ibid., p. 183.
[8] Ibid., p. 186
[9] Ibid., p. 188.
[10] Ibid., p. 191.
[11] Ibid., p. 196
[12] Ibid., p. 197.
[13] Ibid., p. 198.

It is possible to find philosophers that differ on nearly any point one
wants to make. However, those that start with the Greeks and continue to
the Scholastics in working out the truth of reality, one finds commonality.

> Substance is a potency in respect to accidents it can receive. An
> essence is a potency to the act of existance it receives. . . . All these
> are mixed potencies; that is, they are actual in some respect, but in
> potency in another respect. Primary matter, on the other hand is
> pure potency. It is actuated by a substantial form, and with it con-
> stitutes a substance. [14]

Notice that in the quote above "An essence is a potency to the act of
existance it receives. . . ." Time has the parts of future, present and past.
Here we are interested primarily in the future and present. In the above
quote "An essence is a potency to the act of existance it receives. . . ." Or
the future is in potency to the act of the present it receives.

> . . . an act can receive a further act to which it is in potency, as a
> substance which receives accidents—for example, water is actual,
> but in potency of heat that it can receive. But to assert that the
> same thing can be potency and act in the same respect is to assert
> the contradictory. It would mean saying that to be heatable (not
> hot) and to be hot are identical. . . .
> It follows that two intrinsic principles are needed to account for
> any finite thing: a principle of act or perfection to explain its actu-
> ality, and a principle of potency to explain its limitations and be-
> coming. No act is found to be limited save by potency. To say
> something is high in the scale of being is to say that it has greater
> actuality than lesser entities. [15]

From this we see that cold water is in potency of being hot. Hot water
is not; it already has that perfection. Perfection can be lost as well as
gained—cold water to hot to cold.

The future is in potency of becoming the perfection called the present
and then the perfection is lost to the past—cold to hot to cold. But, the
future which is in potency to the present cannot become the present except by the influence of something that is already in act (actual), namely

---

[14] Ibid., Young, p. 223.
[15] Ibid., pp. 224-5.

the present— *potency cannot be raised to the level of act by itself, but only by something already in act.* And, the awareness of man is precisely that present—that something—that is in act. That means that since man's awareness is the *act* that brings the potency of the future into being as the present, man necessarily makes the present happen and in so doing regulates how fast this happens by his collective awareness.

Chapter 13

The Main Elements of Our Existance

Summary of Where We Are

One could say that Aristotle, St. Augustine, and St. Thomas had some of the best minds the human race has produced. All three agree that there is no past and no future because the past in no more and the future is not yet. They are ambiguous about the present other than to say is has no length. In addition Aristotle and St. Thomas say that time is made up of two elements, one is motion of real things in the real world that exists apart from us, and the other is the numbering of motion that can only be done by intelligence emanating from a spirit namely our souls. St. Thomas goes a step further and says time is a quasi substance by saying it is composed of matter—past and future—and form—the soul's activity of numbering. Modern philosophers add little to this. As we have seen earlier, Gardeil offers a quote from St. Thomas, ". . . time, it seems, can hardly exist without a mind to piece the parts together. . . . that is why the Philosopher says that without a soul there would be no time." [1] In addition he says "The idea that cosmic motion is a unitary system, or that a regulating principle of time is necessary—these if not others, are far from gone." [2]

There will be no attempt to argue with the above conclusions. It does seem a bit odd, though, as far as this author could find that none of these experts seems to have taken the subject any further. That last quote from Gardeil implies that it is the cosmos that regulates time. Why would one assume that the regulation of time is done by matter rather than a bunch

---

[1] Ibid., Gardeil, Vol. II, Cosmology, p. 124.
[2] Ibid., p. 126.

of intelligent, powerful spirits that are connected to that matter, that is, the human race?

Now, it is time to take a further step to see how the physical universe, human spirits, human bodies and time are connected.

Time vs. Energy

When we say time started with man and will end with him, we must include the physical universe as well, because at the end of time there will be "a new heaven and a new earth." As the ancients have said, time is the measure of motion and without a flow of energy there is no motion. So, not only do people control the flow of time, they keep the universe in existance as a secondary cause after divine providence. More on that later.

We could postulate that time might be considered as the flow of energy or at least closely related to energy's flow. The future is the potential energy in the universe, the present is that cusp where some of the potential energy is converted to kinetic energy producing motion, and the past is the spent energy which is called entropy. The instantaneous consumption of potential energy in the form of kinetic energy is what makes all things move hence the premise that the flow of time and the flow of energy are connected. This is not such a profound statement seeing that all thoughtful people consider time the numbering of motion, and motion, or more precisely the change of motion, is the consumption of energy. In fact it is the heart of the subject we call physics.

Potential energy is in potency to becoming kinetic energy and then to entropy. Kinetic energy cannot simply go back to being potential energy and neither, strictly speaking, can cold water go back to being the identically cold water after being heated because energy has been lost to entropy. In this light, kinetic energy could be called *actual energy* or *real energy* as in *real time* as opposed to potential energy (the future) at one end or entropy (the past) at the other.

A distinction must be made in this regard. A body in uniform motion that is not changing either speed or direction does not consume energy. But such a condition is in reality only a theoretical construct because even a rock in interstellar space will be acted upon by the gravity of the nearest galaxy and thereby be constantly changing speed and direction.

So much said. But, it would be wrong to say time was identically equal to energy. Energy is part of the physical universe and as we have seen energy and matter are convertible one to the other. Time is strictly speaking associated with the human spirit. This would mean that time enters into the *control* of the flow of energy while not being part of it.

In addition we must be careful not to fall into the existential mindset of "I think, therefore I am" which says that the physical world out there is nothing but a projection of the mind. The physical world is really out there aside from me. However, as stated before, without rational, aware human beings, there would not be much point of any of it.

It has also been mentioned that evolutionists would say that the natural timing mechanisms of nature—movement of astronomical bodies, vibrations of subatomic particles—were there before us and we evolved using those rhythms. But, as we have seen a good case can be made for man as the controlling element for the flow of time. That means those natural rhythms have been dictated by the speed of flow of time we like.

The Structure of Matter

Matter will be discussed in what follows so it is well to see what modern physics can tell us about the subject. Taking the simplest atom, that of hydrogen, we see that its nucleus is made up of a single particle, a proton, and it has one electron "orbiting" around it. Putting the size of these particles and that atom on a scale we can comprehend, we see that if the proton is the size of a B-B and it is placed on home plate of a baseball field, the electron is smaller than a spec of dust in far center field. All the rest is empty space. But it is not entirely empty. It is filled with the force fields caused by the two particles. The force is transmitted by "force carrier particles" such as photons, virtual particles, and gluons. Photons we know about because they are light waves. The other two are something like mathematical constructs nuclear physicists use to make sense of subatomic reactions. As such, they cannot be observed directly because any such attempt changes the processes in which they would be expected to appear.

Nuclear physicists no longer think of an atom like hydrogen as a proton with an electron orbiting around it like a planet about a star. Now, due to quantum mechanics that atom is thought of as a proton with the location of the electron represented by a "cloud" surrounding the nucleus. The cloud is the force field created by the motion of the electron which travels at about two million meters per second which is about a hundredth the speed of light.

Here is it well to think of the hydrogen atom as it really is. Above we compared it to a baseball field to get as sense of its size relative to the size of the proton and the electron but it certainly is not that large. In fact it is very small. It's diameter is 0.1 nano meters or 0.1×10^{-9} meters. If it is that small and the electron is traveling at about a hundredth the speed of light it orbits the nucleus about 6×10^{17} times a second. [3] At that rate it would appear to be practically everywhere on the sphere around the nucleus all the time thus giving a sense of a solid object though since hydrogen is a gas we would not be aware of it.

However all of the other atoms are similar to hydrogen, though with increasingly more complicated structures as one goes up the periodic chart of the elements. This means that all matter contains very little in the way of "stuff" we can actually identify, and a solid wall is not solid at all, but a web of force fields created by the electrons spinning around the nucleus. It is obvious that if the electrons stopped moving the "cloud" of force surrounding the nuclei of atoms caused by the motion of the electrons would disappear and with it matter as we know it.

Continuing this line of thought we see that modern physics has quantized everything, and that is what is called quantum mechanics. Electromagnetic waves have been shown experimentally to have particle properties. As mentioned above these are called photons and they carry differing amounts of energy depending upon their frequencies. All energy is reducible to photons. All of matter can be reduced to molecules, atoms, subatomic particles such as electron, protons, and neutrons and them still farther into smaller particles. There is a persistent effort by scientists to discover that time is reducible to individual packets somewhat fancifully called the "chronon" which is a hypothetical quantum or particle of time.

Body, Soul, Motion and Time

In what follows soul is to be taken as meaning the spirit that is the soul of a living human being, not the soul after death.

Few modern scientists will go any further than saying that time is the numbering of motion. They shy away from things philosophical or out of hand reject philosophy as being mutually exclusive of science. An example of a modern physicist who is willing to discuss philosophy is V. E. Smith in his book *Philosophical Physics*. In fact, he quotes Aristotle and St. Thomas frequently, though in the end he says, with most scientists,

---

[3] These figures are readily available from many sources particularly a text on modern physics, cf.: Kenneth Krane, *Modern Physics*, John Wiley & Sons, Inc., New York, 1983, pp. 133 ff.

that science and philosophy are different realms of knowledge. "The philosophical science of nature does not expect the empiriological physicist to use a philosophical approach in his measurements but only to remember that beyond measurement there is another view of reality which tell us what things are." [4] He, thus, allows for two truths. He states many of the same ideas as do our references above concerning time. In fact, he has a Chapter of 39 pages titled "Time: The Measure of Motion." He has little that is new in all those pages stating again and again the definition of the ancients namely that "Time is the number of motion, according to a *before* and *after*." Emphasis his. [5]

At one point he may have inadvertently fallen into something rather profound for a scientist to say, though as we have seen not at all unusual for philosophers.

> But the real *now* as opposed to the *nows* of reason is the factor in the philosophical physics of time that must receive the heaviest accent. In the real and precise sense, the *now* is a moving boundary, like a point originating or terminating a portion of a line, and when taken in this terminal sense as a limit, the *now* is not a part of time because it is indivisible. Emphasis his. [6]

The part of interest is "the *now* is not a part of time because it is indivisible." First of all the now can hardly be excluded from time because it is the only real part of time there is; it is the part in which we live. But more importantly is his conclusion that the *now* is indivisible. In the references cited earlier the general conclusion is that the present has no length. But, to say the *now* is indivisible, which something that has no dimensions would logically be, implies a little different meaning. It suggests a connection to another thing that is indivisible, namely a spirit.

In the short section on spirit above we saw that a spirit is simple which means it has no parts. It is, therefore as we have observed, indivisible. Human beings are the only creatures that live in the *now* of time. Therefore, they live in this indivisible instant. And *only spirits who are indivisible themselves could live in such a place*. But place is the wrong word because place implies extension and hence has parts—front/back,

---

[4] Vincent Edward Smith, *Philosophical Physics*, New York, Harper Brothers, 1950, p. 393.

[5] Ibid., p. 365.

[6] Ibid. p. 368.

top/bottom. It is reasonable that spirits live in a "place" or rather a "condition" that is indivisible.

One can see how an indivisible now could be associated with and indivisible spirit, but where does that leave the body which is extended in space? We are each a substance composed of a material body and a spiritual soul. As such the soul is the form for the formless matter of the body. As with all substances it is the form that makes the primal matter into a thing. In our case that which constitutes the form is a spirit making humans metaphysically different from all other substances. And, even though the spirit has no parts it animates the entire body to the last cell. In that way our spirits control matter by keeping our bodies alive. In addition they control the motion of physical matter when we perform conscious actions as mentioned above with regard to triune matter.

In our earthly lives the present is ever changing. For pure spirits, if time is a part of them, the present is not changing. In the case of God, He always existed in an unchangeable perfect form. The angels may have had a condition immediately after they were created when time for them was changing and in that time they were given the chance to accept God, or reject Him. After that, time froze for them and they were eternally good or bad. Since God made us in His image and likeness we could assume that means we have time built into us in a fashion somewhat similar to Him and the angels. Only for us, the present keeps changing as long as we live, but will eventually be stopped in eternal bliss or eternal pain where time will be the same for us as it is for God and the angels.

It should be mentioned for completeness that there it the intermediate stage of purgatory. All books on purgatory mention it as extended in time, though a time that flows, or seems to flow to those there, at a different rate from that which we experience. The following account is one of several reported in the literature about purgatory.

> Two religious of eminent virtue vied with each other in leading a holy life. One of them fell sick and learned in a vision that he would soon die and remain in Purgatory only until the first Mass should be celebrated for the repose of his soul. Full of joy at these tidings, he hastened to import them to his friend, and entreated him not to delay the celebration of the Mass which was to open Heaven to him.
>
> He died the following morning and his holy companion lost no time in celebrating the Holy Sacrifice. After Mass, whilst he was

making his thanksgiving, and still continuing to pray for his departed friend, the latter appeared to him radiant with glory, but in a tone sweetly plaintive he asked why the one Mass which he stood in need had been so long delayed. "My blessed brother," replied the Religious, "I delayed so long, you say? I do not understand you." "What! Did you not leave me to suffer for more than a year before offering Mass for the repose of my soul?" "Indeed, my dear brother, I commenced Mass immediately after your death, not a quarter of an hour has elapsed." [7]

We return now to the concepts of potency and act. Recall that only a thing in act can bring a thing in potency into act. And the agent bringing the potency into act must be proportioned to the effect that comes about. For example, a gallon of water at room temperature is in potency of being in act as boiling water. A red hot paper clip is in act of being hotter than boiling water and it contains heat which is the commodity that is required to heat the water. But, its heat is not proportioned to the heat needed to bringing a whole gallon of water to the boiling point.

Looking at the present as time in act and the future as time in potency we see that the present has what is needed to bring the future into the present somewhat analogous to the heat in the paper clip. But is it proportioned to what is needed to bring the potency of the future into the act of the present? Or as in the case where we postulated that time is the flow of energy, is the kinetic energy of the present proportioned to bring the potential energy of the future into kinetic energy. One could say no because there is no real connection between potential energy and kinetic energy as far as the latter causing the former to become kinetic. It is like the water that is still above the dam in relation to the water that is in the process of falling having tipped over the rim of the dam. The water that is falling has lost all connection to the water still above the dam. For the water still above the dam it requires something else besides the falling water to cause it to fall like, for example, the flow of a current pushing it to the rim of the dam and then the assistance of gravity. But, the water already falling is not what causes more water to fall.

That would mean that the present, as such, has no connection with the future. It requires something beyond time to cause some part of that supply of future to become the present. That something beyond time is a

---

[7] Fr. F. X. Schouppe, S.J., *Purgatory*, Rockford, Illinois, Tan Books and Publishing, Inc., 1973, p. 64. First published 1926.

spirit, and the only spirit that is intimately connected to both matter and time is the human soul.

To recapitulate, in his section on time H. D. Gardeil is either quoting Aristotle and St. Thomas Aquinas directly or is summarizing their thoughts. "Though time is not motion it is nevertheless unseverable from motion. Take away all change and motion and time disappears." [8] In addition, "Aristotle allows that time in its full extension cannot exist apart from mind." [9] And again, "And that is why the Philosopher says that without a soul there would be no time." [10]

We can now make a reasonable further statement. Without a soul there is no time so with no supply of time to draw upon there could be no motion. And, with no motion there would not likely be any matter. With all solid matter being little more than force fields made up of virtual particles, as discussed above, if these particles were in some way nullified or their motion stopped the force fields they create would disappear and the universe with them. That is why we can say that the human soul, motion and time are all mutually necessary. Take away the soul and you take away time. Take away time and motion is not possible. Take away motion and matter goes away and with it the universe—*no soul no time; no time no motion; no motion no matter; no matter no universe*. This goes to a previous point that said the universe does seem to be here specifically for us.

The above paragraph states the case negatively. Speaking more positively we can say that we know of only two exemplars of existing things that have no dimensions, the indivisible present and the indivisible spirit. These have what could be called a mode of existance that is different from matter. A dimensionless present cannot control matter that is spread out in space; it is not proportioned to the task. It requires a dimensionless thing that has been created specifically for the task of interacting with, and moving, spread out matter. That special thing is the human soul. Having the capability of associating intimately with matter, the soul necessarily must extend its powers to the other entity that makes the changing world of matter possible. That is, of course, time.

As mentioned before, if a person decides in the morning to go shopping in the afternoon it makes a different set of futures coming toward him than if he stays home and cleans his closet. But, it makes the future

---

[8] Ibid., Gardeil Vol. II, Cosmology, p. 121.
[9] Ibid., p. 123.
[10] Ibid., p. 124.

different for everyone else, too. Certainly the clerk at the store where he might have shopped will have a different afternoon depending on the choice. And, as we have seen the idea of six degrees of separation means that everyone on earth will be affected. Now, all the inhabitants of earth are making similar decisions all the time. This combination of willful, conscious acts is what makes the present and hence the world in which we live. It would not matter if there were only one human in existance, the effect would be the same. And, no, a squirrel grabbing one nut rather than another is not a conscious act.

Final Step

There is a final step that must be taken. That is to ask the question that if time is here for man, and man controls its flow, how does he from moment to moment change that future waiting to happen into the present where he has his existance? Certainly the power of a spirit is capable of such an act, but how is it shaped, how is it made into the continuum of the present? There are at least two factors to consider. One is the spirit of each person alive and the other is Divine Providence. We each make our own present out of the future that is available by our conscious acts, whereas the flow of time and the control of the universe is done with the collective pressure of everyone else and the omniscience of God.

Here the reader is encouraged to go back and read the short section on triune matter. If the soul can affect matter in the brain so we can move our muscles at the command of our mind, it is reasonable the collective spirits of all who are alive can move the amalgam of matter in the universe in a way suitable to the human race. It is reasonable to say that is what we call time. It would seem that a continuous flow of time requires many people because it flows along when we sleep or even daydream and make no decisions. Yet, as long as we are alive our individual spirits must keep us *in* the present whether or not we are conscious of it. It may be that is the main physical—for lack of a better word—duty of our souls vis-à-vis our individual selves. If we are kept in step with the present it would mean all of our other physical processes such as breathing and heart beat are kept functioning. We finally come to a point where we can form a definition of time.

Time is an accident (attribute) of man's spiritual soul that exhibits itself when the soul is united with a living body. Its purpose is to keep the physical universe in existance and with it man.

We started by postulating that the essence of time was an accident of human beings that are alive. And that "An accident is an essence whose nature it is to exist in a substance as in a subject." [11] In addition St. Thomas leaves intact Aristotle's assertion that "without a soul there would be no time" [12] so this definition is not something new but a way of restating long accepted philosophical concepts.

The reader my object to the last statement of the definition on the grounds that we have all been told the universe is kept in existence by divine providence. That is a true statement, but divine providence also makes use of secondary causes. "The truth that God is at work in all the actions of his creatures is inseparable from faith in God the creator. God is the first cause who operates in and though secondary causes. . . ." [13]

The orderly movements of the planets around the sun are caused by divine providence but we recognize a secondary cause at work, namely gravity. Certainly God could make that happen without gravity but it would leave things not making sense. Likewise, when one starts his car and presses on the accelerator the car moves because the tires do not slip on the pavement—another case of divine providence and yet another example of a secondary cause, namely friction. As seems to be His way, God has made the universe in such a way that if we look long enough we will find an explanation of how things work. It is not unreasonable to say that man as a secondary cause keeps the universe in existence—no man, no time; no time, no motion; no motion no matter; no matter no universe.

There is yet another issue about time and the universe to be mentioned. We start by looking at a sheet of normal copy paper. In the United States that has a width and height of 8-1/2 by 11 inches and it's about 0.004 inches thick. What happens to the width and height of the paper when the thickness is reduced to zero? The sheet of paper disappears and with it width and length. We have already arrived at the conclusion that the present—where the universe exists—has no length. Is that like reducing the thickness of the paper to zero? It would imply that the universe doesn't exist other than by having an infinite number of infinity short presents pieced together by the human spirit. Infinity is a hard word normally attributed only to God and maybe that's a place there divine providence especially comes into play.

---

[11] Ibid., Thomas Aquinas, *Thomas Aquinas On Being and Essence*, p. 66.

[12] Ibid., H.D. Gardeil, O. P. *Introduction to the Philosophy of St. Thomas Aquinas*, Vol. II, pp. 123-4.

[13] Catechism of the Catholic Church, 1994, United States Catholic Conference, Inc.—Libreria Editrice Vaticana §308.

To think of this another way consider that we accept as true that a human being is a substance composed of a material body and a spiritual soul and that's that. No one ever seems to question how two so dissimilar "things" could come together. It is normally left to something like God knows what He's doing so forget about it. Let's be a little discourteous and ask the question. The spirit has no parts and the present being indivisible has no parts either; and the present is where we and the universe exists. It seems not only possible but entirely expected that the human spirit and the partless time/universe are specifically made for each other and naturally come together although with the spirit being the superior member.

There remains the little problem that the human spirit and only a small part of the universe come together in the human being. That leaves the rest of the universe in the hands of God. We must think back to where we considered that after the general judgement there will be a new heaven and new earth and the present earth will have passed away. With no more people like we are now, either God withdraws divine providence from the present earth or people as a secondary cause leave the present earth unattended and it disappears. This would mean that in some way the union of spirit and body has an extended effect on the universe. And as we have shown before, spirits certainly have the power to do that. It might be well to go back and reread the section in spirit in this regard especially the part about the power of the spirit.

If the reader finds all of this a bit much there are several facts, some of which are listed below, that must be taken into account if our world and our lives are to make sense.

- Human beings are made up of a spiritual soul and a material body.
- Those two things are from two different realms of being.
- Humans live in time along with the rest of the universe.
- Time is thought of as having three parts—past, present and future.
- The future does not exist because it is not yet.
- The past does not exist because it in no more.
- That leaves the present as the only part of time that exists.
- The present has that annoying property of having no length—indivisible.
- The spirit is simple in that it has no parts—also indivisible.

The Fossil Record Revisited

The fossil record really has nothing to do with time. Here will be presented a somewhat whimsical explanation of why we have it. It is included since our whole lives have been permeated with the billions and billions of years needed for the universe and us to "evolve." To have that suddenly taken away may leave some readers feeling a bit quaky inside. So, if there were no time and no universe before the first man and woman, why is there a fossil record? In a simple sentence, the fossil record was placed in the earth as a temptation to man. Is that a non-meaningful statement? Not really. The reasoning goes like this.

In Genesis we read that God created the universe and man in six days. The evolutionist will say that if someone wants to hold that position, fine, but the six days are not six literal days as we know them, but six epochs. Divide up the thirteen billion years since the big bang into six chunks of whatever length suits your fancy. Others insist that the six days were really six days as we now know them. Why not, they say, God is all powerful and can do anything he wants. It's difficult to argue with that.

There is a third way, though, and that starts with God deciding to make man composed of a material body and a spiritual soul. Having decided that, he naturally had to provide a place for him to live, that is, provide the *demands of nature*, a subject already covered. That being the case, in a blink of His eye there was the universe complete with the earth containing the Garden of Eden and the fossil record. To make that concept understandable to the people of his time Moses used the metaphor of six days in the first chapter of Genesis.

The first man and woman were placed in the garden oblivious of the fossil record and were told they could live in happiness and contentment with no scorchingly hot days or blizzards, no hunger, no pain. They would work but work would be enjoyable. If they became tired before the work was complete they could simply call upon some of the preternatural gifts they possessed and complete the work by a sort of psychokinesis with no physical effort on their part. After a certain period in the garden they could proceed on to heavenly bliss without suffering death. The only proviso was that there was the somewhat innocuous tree in the center of the garden, the fruit of which they were not to eat. That was the only downside to paradise.

Leave it to a woman, but you guessed it, along came a wondering snake and she took up a conversation with it. In due course both the woman and her mate ate of the forbidden fruit and in consequence were thrown out of paradise. The take away from that, girls, is don't talk to snakes.

Notice that, though the fruit of the forbidden tree was appealing, Eve was not drawn to eat of it out of hunger. There were all the other trees of which they could eat. However, to vastly compress Christian theology, there was after the fall a way left for them to get to heaven. That was to work by the sweat of their brows while leading virtuous lives, die, atone for sins not yet atoned for and then ascend to heavenly bliss for eternity. Through the ensuing millennia after their expulsion there was a running battle with hunger and, in fact, extinction. They mourned the loss of paradise, and gave not a thought to the fossil record except for a philosopher here and there, but the concept had no effect on society as a whole.

Eventually the collective knowledge of the race of humans arrived at what could be called a technological state. One could say it started with Copernicus and Galileo in the sixteenth and seventeenth centuries, or with Newton in the eighteenth. But from there on, the running battle with hunger began to lessen. Technology from the breeding of plants and animals to advanced farm implements made life more secure. In fact, man was on his way to recreating the paradise he had lost and be free of want and privation. But, God is no dummy—omniscience precludes that—so he placed a time bomb in the earth, namely the fossil record. As seems to be His way, God demands that there is always a decision to be placed before man—to accept God or to reject God.

Along with technology in practical areas of making life easier and better, came knowledge in the life sciences. That in itself was not bad, but it inexorably had to lead to people wondering where life came from and, though it took a leap of faith, the fossil record was made to appear as a map to the past all the way to where the first life popped into existence from non-life.

However, if we as a race had kept our eye on the ball, technology would have left free time for the average person to reflect on this life and the life hereafter and use his leisure time to improve his chances of success vis-à-vis heaven or hell. Even in view of our fallen nature if we maintained a knowledge of God a sense of shame and guilt would have remained. But, the temptation inserted into the created world by the fossil record came to the fore. It made it possible to explain away God and, hence, creation by God and replace it with the flow of evolution in which man was becoming ever more perfected with each passing generation, obvious facts no withstanding. The normal fickleness of man by itself would not have been sufficient to cause the mass apostasy as has happened; it took a general concept that could be made to look as natural as breathing. Enter evolution based on the fossil record.

That is the temptation posed by the fossil record. At the point were the human race could be living in a relatively comfortable state under divine guidance, modern man ate of the forbidden fruit in his oh, so, modern garden.

The breach between God and man caused by eating the modern fruit of evolution is not in a strict sense the same as the original breach. Man has been redeemed and still maintains free will today as he did, say, a thousand years ago so any individual is capable of pulling himself at least partially out of this present fallen state with God's help. The difficulty lies in a type of mass amnesia that leaves people unable to think. It is no idle musing to wonder if it is possible for the human race as a whole to shuffle off the coil of narcissistic atheism which has been poured into the psyche of modern man by the incessant barrage of the evolutionists.

There are many who will object to the above thesis so it is reasonable to ask if the dinosaurs really did exist. The direct answer is that dinosaurs most likely did live but it would have to have been while people were alive. The idea that the universe is thirteen billion years old is based on the big bang theory which says that it all started from a small speck of super dense energy. The explosion sent the primitive universe out in all directions such that at any chosen point it would appear that everything was moving away from that point. All of this is supposedly determined by the light from distant objects being red shifted in proportion to their distance from the observer as a result of the Doppler effect. Evolutionists assume it took thirteen billion years because a lot of time is needed for evolution. However, there are other explanations for the red shift and there is much evidence for a younger and smaller universe than is commonly depicted, the elucidation of which, is beyond our present scope.

With that said, it is reasonable that when the first rational human beings appeared dinosaurs roamed the earth. One example of a young earth which would make that possible is pieces of fired pottery and gold jewelry that were found in a coal seam supposedly thirty million years old. These artifacts were present when the coal was formed; they were not interjected later. Another example is the soft tissue found in dinosaur bones that are supposedly millions of years old. There are many other such examples, but the evolutionists reject them as outliers. They are not outliers, they are facts that must be accommodated to any sane theory.

PART III

LORDS OF THE WORLD

Chapter 14

Their History

Introduction

It may seen that the title of this part is a bit odd but the lords of the world are the Jews as we shall see. We will start with a short history of the Jews and then various specific topics will be expanded upon concerning them at the present time. Here again, as in the part on evolution and then on time, it is important to view many aspects of the subject so the reader has a chance to think about yet another area that is rarely seen in print, spoken about or even thought about. We will see why this is the case and venture into areas that are nothing short of fascinating, though, hardly hidden or unknown to one who takes the time to make inquiry.

It is not the case that this section is meant to inveigh against the Jews as a specific target. It is just that while they make up only two tenths of one percent of the worlds population they effectively control the whole of mankind intellectually, politically and economically. As such, it is not anti-Semitism to simply point out what their role is in the world society. Further, if the reader objects to this or that point presented below as being questionable it must be borne in mind that opinion on most of the subjects varies considerably. Even among Jewish writers there is a wide divergence of what are and are not the facts on certain subjects and especially what the facts mean.

Before we go further, it is necessary to define what one means when he speaks of the Jews and it is not easy. In the Catholic Church it has historically meant those members of the Jewish race or others who hold some or all of the tenants of the Tanakh (more on that later), and generally were of the race that is described in what Christians call the Old Testament, but have rejected Christ. In the middle ages when a Jew converted to Catholicism he ceased to be a Jew in the eyes of the Church and became a Catholic in spite of his race. Statements such as, "They hid for fear of the Jews," are seen in scripture. It does not say all the Jews, of course, because Jesus and the apostles were Jews themselves. And it does not say a few of the Jews. It simply says *The Jews*.

In the New Testament the words Jew or Jews are almost always used in a derogatory sense. The main exception is where Jesus is talking to the Samaritan woman where he says, ". . . for salvation is from the Jews," Jn 4:22. Otherwise when referring to the chosen people, including Jesus' followers, the words Israelite or Hebrew are used. In writing his letter to his countrymen living in the holy land St. Paul addresses it to the Hebrews not the Jews.

To be more specific, when we say the Jews we do not mean all Jews, but the Jewish people organized some way as under the Leadership Conference of Presidents of Major Jewish Organizations which includes 51 Jewish organizations in the United States. Most of the members of these organizations are manipulated by the leaders much as they were at the time of Christ. That does not change the fact that they make up a body which acts as a body. This is the way the New Testament uses the term "The Jews." It does not say all Jews, nor does it say Jewish leaders.

To use an example from recent secular history we say the Japanese bombed Pearl Harbor on December 7, 1941, which was the specific cause for the United States entering World War II. We do not say Japanese pilots bombed Pearl Harbor, nor to we say the Japanese leaders did it. We mean the Japanese people organized as a body, as a nation, did it. Certainly neither the Australians nor, in fact, the Germans bombed Pearl Harbor, the Japanese did.

A further comment must be made in this regard. When many people hear the above statements for the first time they will disagree and say something like, "Well, I know some Jews and they're good, honest people," as if that ends the discussion by refuting the argument. If you want do avoid overt hostilities do not point out to them that what they just said is like saying, "I know a man with one arm, therefore all men have one arm." This is one small example of the end of thought. Further, it is hoped that in what follows the reader does not take away the idea that there are not good, honest Jews. Most Jews are good, honest people just like those of any other race.

One must not forget that after the destruction of the temple in Jerusalem in AD 70 the Jews became a rootless people. The fact that there is a state of Israel today does little to change that. Even before the destruction of the temple the Jewish people were widely dispersed. On Pentecost, we see that there were Jews in Jerusalem from all over the then known world, Acts 2:5. As a result of being rootless, they were forced to live in other countries and cultures. They are always aware their rootlessness will make them stand out so they set about uprooting the host society because if everybody is rootless they will be less noticed. The difficulty with that is no rootless society lasts for long. To put it another way, the Jews will frequently cause disruptions in their host society so the people have bigger things to worry about than the Jews.

Jews tend to stick together more than other identifiable groups of people. For example, the Ivy League universities have about thirty percent of the student body made up of Jews while Jews represent two to three percent of the U.S. population. The admission policies favor Jews in a number of ways. Young Jews will always accept the privileges their race provides them. Then, when the Jew graduates, say, from Harvard, he will be offered a good position by other Jews. Later when he is asked to do something to help a Jewish cause he has no alternative but to do as he is told. In this way, even if most Jews are not actively involved in destabilizing their host societies, nearly all will lend a hand when asked to do so.

This cohesiveness is seen in the way they seek out others. When a Jew meets another person that they suspect of being a Jew they frequently will quietly ask, "M-O-T?" They are asking "Are you a Member Of the Tribe?" that is, the tribe of Israel hence the question is about ones race not religion. It is probably more frequently used negatively as in, "That person is not M-O-T."

We will see in the section on Jewish Demographics that only twenty-three percent of Jews in the U.S. go to Synagogue once or twice a month. Yet, all Jews are aware of their ethnicity in a way no other race is. Even though gentiles seldom think of it that way, there is no avoiding it. Jews, and that includes every one, are aware that the Catholic religion comes from its Jewish roots in present day Israel. Islam is a major religion that is essentially a Catholic heresy as is Protestantism. Those three main religions include at least half of humanity. The Jews know who they are, and what they are has had a significant part on world history for the last several thousand years.

The History

In the Old Testament, there were many prophecies of the coming Messiah, but they were of two types, the Messiah "bar David," and the Messiah "bar Jonah." Here "bar" means "son of." The Messiah bar David was to be a worldly leader, a king, a general. The Messiah bar Jonah would be a spiritual leader. As an example of the earthly king idea look at Psalm 149, the second part of which reads as follows.

> And let two-edged swords be in their hands:
> To execute vengeance on the nations,
>   punishments on the peoples;
> To bind their kings with chains,
>   their nobles with fetters of iron;
> To execute on them the written sentence,
> This is the glory of the faithful.

This is a far piece from turn the other cheek; do good to those who persecute you; pray for your enemies. The Messiah bar Jonah is most clearly seen in the book of Jonah which will be discussed later where we will go into what the Jews understand the Messiah to be. As an aside Messiah is the Hebrew word for anointed one, and Christ is the Greek word for the same thing.

At the time of the coming of Christ, the Jews were in a particularly bad way having been under the heel of the Roman Empire for forty years. They were suffering and the prophecies said the Messiah would be coming soon. As a result, they longed for the coming of the Messiah who just *had* to be bar David. The Messiah came and guess what? He was bar Jonah, a spiritual savior. This is one of the primarily reasons the Jews, the leaders particularly, rejected Jesus.

In Matthew 16:17 Jesus asks the apostles who the Son of Man is. They gave various answers but Peter says, "Thou art the Messiah, the

Son of the living God." And Jesus answers, "Blessed art thou, Simon bar Jonah. Flesh and blood have not revealed this to you, but my Father who is in heaven." It appears accepted by the early church fathers that Peter's biological father was named John or Johanan. Some say that Jonah is simply a shortened version of Johanan like we use Pete for Peter today. This is born out in John 1:42 and 21:16-17 where most translations use "son of John" for at least one of those citations. However, it is interesting that Jesus made the statement in that form because it brings into play bar Jonah as opposed to bar David. In recent American translations of the Bible, the passage is rendered ". . . Simon son of Jonah." In the Douay Rheims version of the Bible it uses "bar" for "son of." The change from the use of "bar" to "son of" in this passage deprives the reader of the important connection between Jesus and the Messiah bar Jonah. It should be noted that the Hebrew word for "son" or "son of" is ben, so in what follows ben will frequently be used especially when quoting Jewish sources.

The use of "bar" as opposed to "son of" is brought out in Pope Benedict XVI's book *Jesus of Nazareth*, where he discusses the Jew's demanding Barabbas over Jesus.

> . . . For the fusion of faith and political power always comes at a price: faith becomes the servant of power and must bend to its criteria.
>
> The alternative that is at stake here appears in a dramatic form in the narrative of the Lord's passion. At the culmination of Jesus' trial, Pilate presents the people with a choice between Jesus and Barabbas. One of the two will be released. But who was Barabbas? "Barabbas was a robber" (Jn 18:40). But the Greek word for "robber" had acquired a specific meaning in the political situation that obtained at the time in Palestine. It had become a synonym for "resistance fighter." Barabbas had taken part in an uprising (cf. Mk 15:7), and furthermore—in that context—had been accused of murder (cf. Lk 23:19, 25). When Matthew remarks that Barabbas was "a notorious prisoner" (Mt 27:16), this is evidence that he was one of the prominent resistive fighters, in fact probably the actual leader of that particular uprising.
>
> In other words, Barabbas was a messianic figure. The choice of

Jesus versus Barabbas is not accidental; two Messiah figures, two forms of messianic belief stand in opposition. This becomes even clearer when we consider that the name Bar-Abbas means "son of the father." This is a typical messianic appellation, the cultic name of the prominent leader of the messianic movement. The last great Jewish messianic war was fought in the year A.D. 132 by Bar-Kokbba, "son of the star." The form of the name is the same, and it stands for the same intention.

Origen, a Father of the Church, provides us with another interesting detail. Up until the third century, many manuscripts of the Gospels referred to the man in question here as "Jesus Barabbas"—"Jesus son of the father." Barabbas figures here as a sort of alter ego of Jesus, who makes the same claim but understands it in a completely different way. So the choice is between a Messiah who leads an armed struggle, promises freedom and a kingdom of one's own, and this mysterious Jesus who proclaims that losing oneself is the way to life. Is it any wonder that the crowds preferred Barabbas? . . .

If we were to choose today, would Jesus of Nazareth, the Son of Mary, the Son of the Father, have a chance? [1]

On Pentecost, not called that yet of course, every single Catholic in the world was a Jew, or more appropriately a Hebrew or Israelite. It did not take too long before almost no Catholics were of that race. The chosen people had missed the boat. Over time the Jews split into two groups, those who still looked for the coming of the Messiah (presumably bar David), and those who gave up. The first group are commonly called the religious or observant Jews, and the second the secular Jews. It is important to note that Jews are a race *and* a religion. The two are almost identically equal. This is unique in the world, in all history. As a result, the Jews who gave up did not simply fade away to be intermarried with other races and effectively disappear. They had their own plan.

Keep in mind that Catholics are particularly hated by most of the Jews. Catholics have the audacity to say that after waiting for nearly two

---

[1] Pope Benedict XVI, *Jesus of Nazareth*, Part 1, New York, Doubleday, 2007, pp. 40-41.

thousand years the Jews were so stupid that when the Messiah came they missed him and the hated gentiles accepted him. If that were not enough, we also say that all people everywhere at all times are saved by Jesus Christ. To add further insult to injury, it is frequently said that the Jews murdered God, their God. Wow! That would anger anybody. This led to an animosity between the Jews and the early Christians that has lasted to this day.

An early example of this was the persecutions in the first centuries. In the couple of centuries either side of the coming of Christ the Roman empire was polytheistic through and through. With each new conquered people they got a new batch of gods. Eventually there was a god for everything, a god for love, Eros, and Mars the god of war as most of us know. On the more mundane side of things there was a god, more correctly goddess, of drains and sewers. Her name was Cloacina. She initially presided over the city of Rome's Cloaca Maxima, the city's primary sewer, and later took on additional duties of guarding the lesser tributaries of the disposal system. Not surprisingly Cloacina enjoyed a robust following especially among home owners. One should not forget Janus the god of portals, of beginnings and endings, from which comes the name of the first month of the year, January, since January marks the end of the old year and the entry into a new one.

With this in mind it seems odd that the Romans would mount severe persecutions against the Christians. After all, what difference did one more god and the associated religious cult make? Certainly the powerful Roman legions could handle one more sect if it became unruly. Add to that was the fact that the Christians were by their creed remarkably peaceful.

In fairness it must be mentioned that ever thought the Christian religion was peaceful it had an iron hard antipathy to the world culture of the time. This hostility came not in any outward actions but in its fundamental make up. "What made Christianity so dangerous was its uncompromising, radical de-divinization of the world." [2] Gone were the gods of thunder, lightning, and volcanoes. We saw this in the section on evolution. What is evolution but a return to a "divine" world in the from of

---

[2] Eric Voegelin, *The New Science of Politics*, Chicago, The University Of Chicago Press, 1952, p. 100.

pantheism.

However, the persecutions would not have been as severe as they were had the Jews not insinuated themselves behind the scenes of the Roman government. This repugnant attitude toward the Christians went all the way back to the Gospel account of the passion in John's Gospel where the Jews tell Pontius Pilate, "If thou release this man thou art no friend of Caesar; for anyone who makes himself king sets himself against Caesar." Jn. 19:12. It is of note that Pilate surely knew that his real danger might have come from not releasing Jesus. Even in the case of the most heinous criminals the trial was held on one day, the sentencing on a second and the execution of the sentence on yet a third. That was Roman law. Here, all three, in an unprecedented fashion, were carried out in a single day.

The Christians survived the persecutions until finally in 313 Emperor Constantine proclaimed the Edict of Toleration whereby the empire was officially neutral toward religion. Constantine died in 337 and was followed by a succession of emperors until in 361 Flavius Claudius Julianus, known as Julian the Apostate to Catholics, became emperor. He was most favorable toward the Jews not so much because he liked Jews, but because he hated Christians. He even went to the point of making an effort to rebuild the temple in Jerusalem which undertaking ended in disaster. However he only ruled two years dying in June of 363. This was perhaps the high point for the Jews because things kept slipping for them until in 380 Emperor Theodosius decreed Christianity to be the official religion of the Roman Empire. "In 17 years, the Jews had suffered a complete reversal of fortune. The people who cried 'We have no King but Caesar' found that Caesar had suddenly become a Christian." [3]

Over the centuries the Jews kept looking for the coming of the Messiah, always of the "bar David" type. There were many false starts until in the seventeenth century a very charismatic man came along who looked like the real thing. May 31, 1665 was a fateful day for the Jews. On that day, Sabbatai Zevi, an Ottoman Jew, who lived from 1626 to 1676, proclaimed himself the Messiah in Gaza and swept with him the whole community, including its rabbis. The rabbis all through Europe

---

[3] E. Michael Jones *The Jewish Revolutionary Spirit*, State Line, PA., Fidelity Press, 2008, p. 79.

were not far behind. Yet today, Sabbatai Zevi is mostly unknown and there is a reason for that.

In 1666 Sabbatai sailed to Constantinople to take the crown from the head of Sultan Mehmet IV, the Ottoman ruler, with the intent of converting the Moslems to Judaism. Before he got to the city, soldiers of the Sultan captured him and put him to prison in Adrianople where the Sultan gave him the choice: either convert to Islam or be put to death. Zevi said, "Well, since you put it that way, I'll become a Moslem."

The shock wave which spread through the European Jewish communities was the biggest catastrophe to hit the Jews since the destruction of the temple in Jerusalem in 70 AD. After this many Jews stopped waiting for the Messiah and started looking for substitutes. In fact, they set about to make the bar David case into reality, with or without God's help. They were tired of waiting. By 1879 Baruch Levy wrote to Karl Marx announcing that henceforth the Jewish people, taken collectively, will be its own Messiah. And, make no mistake about it, this Messiah would in fact be bar David.

Since the time of Christ, the Jews have been at war with the Catholic Church. But after 1879 it seemed they possessed a new and terrible earnestness as the "revolutionary Jews" set about establishing the Jewish people as the earthly rulers, and not just over the state of Israel created in 1948, but the whole world. It seems that after 1879 they developed a unity of purpose not seen before. In various places and times this "bar David" mentality became expressed in violence as they pushed their desire to make heaven on earth, the earthly paradise. This led to widespread atheism among the Jews.

Here will be presented a few examples of that this Jewish fervor meant. In the late nineteenth century, a young Englishman named Cecil John Rhodes (1853-1902) became obscenely wealthy in the South African diamond business. He wanted to use his wealth to do good for mankind, or so he said. He decided that to bring about world peace it was necessary to destroy the existing national boundaries and national cultures that conflicted with one another. And, the quickest and easiest way to change these societal structures was through war. So, the way to peace was through war. He is the source of the money for the "Rhodes Scholar" program. President Bill Clinton was a Rhodes Scholar, remember?

How does this fit into the story? Cecil Rhodes was a British Jew, a student and a devoted fan of another British Jew, John Ruskin (1819-1900). The greatest evidence of collusion between the Illuminati and early Imperialists is the great friendship between this racist colonizer Cecil Rhodes and the house of Rothschild. This has been detailed in Fritz Springmeier's *Bloodlines of the Illuminati*. The Rothschilds, you will recall, are to this day the largest banking conglomerate in the world. The founder of the Rothschild dynasty, Mayer Aschel Rothschild, once was asked why he worked so hard, after all, he had more money than he could ever spend. This may not be stated verbatim, but he replied, "You have no idea how great it is to feel the bones of millions of Christians crunching beneath my boots."

Jews started Communism. Karl Marx was a Jew from what was called East Germany. Eventually a great many in the Russian Revolution were Jews. With the notable exception of Lenin, who was only half Jew, nearly all of the leading Communists who took control of Russia in the period 1917 - 1920 were Jews. Leon Trotsky was a Jew from Brooklyn, New York. Stalin was a Jew, as was Khrushchev who came after him. It is not clear if Vladimir Putin is a Jew but it could be expected because at one time he was the godfather of a Russian Mafia clan as well as heading the KGB. It's hard to tell because Putin's bio on the Internet does not even give the names of his parents.

There were about three percent Jews in Russia in the early part of the twentieth century, about the same percent as there are in the U.S. today. In Russia they made a royal nuisance of themselves as terrorists. They blew up trains, set fires, assassinated public officials, kidnapped rich people for ransom, robbed banks, etc. It must be noted that not all the terrorists were Jews. As it was said, only ten percent of the Jews were terrorists, but fifty percent of the terrorists were Jews. While the Jews played a large role in the Communist Revolution they could not have done it alone, that is, without the hungry Russian peasants. But, it is certain that without the part played by the Jews the Communist Revolution would not have happened.

In the book *The Plot Against The Church* Maurice Pinay details what could be called the long history of the Jews. He has some chapters devoted to the Russian Revolution of 1917. By 1918 when the Communists were firmly entrenched, of the 502 top positions in the new regime, 459 were occupied by Jews. Pinay does not just include this as a statistic, but lists all 502 positions, the name of the occupant, and his race. The book is well researched. As you might expect, a Jew ran Stalin's Gulag prison system, too.

Next we come to the Ukraine in 1932. The Ukraine had had a good harvest. But, Russia was going broke and the people were starving. The starvation was brought upon them by Joseph Stalin imposing the collective method of farming wherein there was no reward for hard work. This resulted in little planting and even less harvesting. The government needed hard currency and food so they decided to appropriate the Ukrainian wheat for their own uses. That meant there would be nothing for the Ukrainians to live on and they would starve to death, but the order was given. It was hard to get Russians to over-see such a devilish plan especially since the Ukrainians were blood brothers to the Russians. As a result Jews were sent to oversee the job. After all, they had already liquidated millions in Russia. This led to the starvation of between ten and thirteen million Ukrainians in the period of 1932-1934. This is known as the Holodomor, The Great Ukrainian Famine or The Ukrainian Holocaust. Say you never heard of that? You're not alone.

Having taken over Russia the communists were terrorizing Germany and surrounding countries during the nineteen twenties and thirties. The Jews were in charge of the terror program. Various countries relied on Jews to do their dirty work because, as in Ukraine they would do things that the indigenous people, no matter how barbaric they were, would not do.

During this time Adolf Hitler was making his appearance. We have this interesting insight into the thinking of Hitler which, at least partially, reveals his animosity toward the Jews. It is a record of a meting between Hitler and the famous theoretical physicist Max Planck.

Following Hitler's seizure of power, I had the responsibility as president of the Kaiser Wilhelm Society of paying my respects to the Fuhrer. I believed I should take this opportunity to put in a favorable word for my Jewish colleague Fritz Haber, without whose invention of the process for producing ammonia from nitrogen in air [which made possible the large scale synthesis of fertilizers and explosives] the previous war [World War I] would have been lost from the start. Hitler answered me with these words: "I have nothing against Jews as such. But Jews are all Communists, and it is the latter who are my enemies; it is against them that my fight is directed." I commented that there are different types of Jews, both worthy and worthless ones to humanity, with old families of the highest German culture among the former; and when I suggested that a distinction would have to be made between them after all, he replied: "That's not right. A Jew is a Jew; all Jews stick together like burrs. Where there is one Jew, all kinds of other Jews gather right away. It would have been the duty of the Jews themselves to draw a dividing line between the various types. They did not do this; and that is why I must act against all Jews equally." He ignored my comment that forcing worthy Jews to emigrate would be equivalent to mutilating ourselves outright, because we direly need their scientific work and their efforts would otherwise go primarily to the benefit of foreign countries. Instead, he broke out in generalities and of the Society." [4]

In this regard it is interesting to speculate about what might have happened if Hitler had taken Planck's advice and kept the "worthy Jews" especially the scientists. In 1939 nuclear fusion was discovered and Enrico Fermi saw the prospect of a chain reaction that could lead to an atomic bomb. But, he fled Italy that same year to the U.S. because of the Nazis. Though, Fermi was not Jewish many on the Manhattan Project that produced the first atomic bomb were. Those physicists on the project that were Jews that came to the U.S. fleeing Nazis included Loe Syilaed, Hans Bethe, Albert Einstein, Niels Bohr and others. With these men at his disposal, Hitler could easily have made atomic bombs first and with them won the war. Notice that Max Planck was pleading on behalf of the Jewish scientists six years before World War II started. He was being a patriotic German. Plots against Hitler did not start until generals and others could see he was losing the war.

Going back to Hitler's concern that the Jews were communists, remember that Germany shared a border with the USSR. He knew of the horrors in that regime even if it was largely kept from the people of the United States. It is not the intent here to say nice things about Hitler; it is

---

[4] Max Planck, 'Mein Besuch bei Adolf Hitler', Physikalische Blatter, Volume 3, 1947, p. 143.

just interesting to know that is what he said.

Leading up to the Second World War, the feeling among the Germans against the Jews had been heightened because they were terrorists, and by the way, they owned and were running most of Germany. "In the year 1918 Germany was showplace of a Communist, Jew directed revolution. The Red councils of the republic of Munich was Jewish . . . . With the fall of the monarchy the Jews gained control of the country and the German government." [5] It is interesting to speculate on the course of European history, and our own, if it had not been for Hitler. It is also interesting to note how thoroughly our history texts have been scrubbed of these important historical facts. In the end, though, the Communists did get half of Germany. However, once the holocaust started Jews began leaving Germany and German occupied lands in droves.

When the time came to find men to operate the Nazi concentration camps, especially those that contained mostly Jews, what did Hitler do? He found survivors of the Ukrainian holocaust to do the job. That was not so hard because they had a big score to settle with Jews. Never heard of that? Once again you're not alone.

Another interesting thing that is little known is that in the Weimar Republic (1919-1933) the idea of eugenics was spreading. Due to the extreme losses of WW I the prospect of supporting the unfit appeared untenable. As will be seen in the following section on the Talmud, the Jews, as a matter of there faith—whether or not most subscribe to it—is the belief that only Jews are fully human. Hence, the killing of non-Jews, fit or not, would not be cause for concern. This idea could not have escaped the Teutonic Germans since by the time Hitler took over Jews owned at least one third of Germany which meant that the feeling against Jews was running strong. "Though there were little more than half a million of them [Jews] in the midst of a people of sixty-two million—less than one percent of the population—their control of the national wealth and power lost all relation to their numbers." [6] It was not a big step in the thinking of the Germans to see the Jews as being unfit to live. Or to say it

---

[5] Maurice Pinay, *The Plot Against The Church*, first published in Italian in 1962, English edition: Palmdale, CA, Christian Book Club of America, 1967, p 41.

[6] Arthur Bryant, *Unfinished Victory*, London, Macmillan & Co. Ltd., 1940, p. 139. This is a good history of the Jews in Germany in the decades before WW II without the coloration of the holocaust that followed.

another way, has anyone ever bothered to ask the question of where the Germans got the idea of a pure Arian race? They got it as much as anything from the Jews.

There is such a thing as the Jewish Mafia. When the Iron Curtain came done in 1990, the main line news reported that the economy in Russia would have disappeared all together if it had not been for the Mafia. That Mafia was and is run by the Jews. It is said that the USSR would have collapsed ten years sooner if it had not been for the Jewish Mafia propping it up. So, if the Jews were running the Russian government, why did they let the Jewish Mafia run wild? Was it a case of the right hand not knowing what the left was doing? Sort of. All over the world it seems that wealthy industrialists and financiers would be dead set against communists. But, if these people are Jews, and most of them are, they are also communists and the purpose of communism is for the leaders to take for themselves that which has been hitherto beyond their grasp. That is, the Jews in one way or another are all in it together even if along the way some Jews perish in the grab for wealth and power.

The international Jewish Mafia is still extensive today. The present head of this world-wide Mafia is Semion Mogilevich. Some say he is the most powerful man in the world. He has his own army, artillery, antiaircraft guns, and missiles. He may even have nuclear weapons from the former Warsaw Pact countries, and is presently trading with various governments to provide them with nuclear technology. NATO has said that he is a threat to the stability of Europe, though his name is little known.

Chapter 15

The Situation Today

Jewish Demographics

Before going further it is well to look at Jewish demographics in the United States. Overall, Jews make up about 2.2 percent of Americans, at an estimated 6.8 million according to a Pew study. By comparison, 6.06 million Jews live in Israel, according to Israel's Central Bureau of Statistics. The study found that about 10 percent of American Jews are former Soviet Jews. These came to the U.S. after the fall of the USSR. It was both because they were now allowed to emigrate and to the fact that the Jewish Mafia in Russia had less power with the fall of communism.

Of the Jews in the United States a growing proportion of them are unlikely to raise their children Jewish or connect with Jewish institutions. Two-thirds of them are not raising their children Jewish at all. Overall, the intermarriage rate is at 58 percent, up from 43 percent in 1990 and 17 percent in 1970. Among non-Orthodox Jews, the intermarriage rate is 71 percent. Overall, 22 percent of U.S. Jews describe themselves as having no religion, yet emotional attachment to Israel has held steady over the last decade, with 69 percent of respondents saying they feel attached or very attached to Israel.

Approximately one-quarter of Jews said religion is very important in their lives with less than one-third of American Jews say they belong to a synagogue. Twenty-three percent of U.S. Jews say they attend synagogue at least once or twice a month.

The survey also asked respondents about what it means to be Jewish, offering several options. The most popular element was remembering the

Holocaust, at 73 percent. [1] The reason for this will be covered later in the section on the Nazi Holocaust.

That last statistic is telling. In spite of the apparent lack of interest in Jewishness among Jews, 73 percent relate to the Holocaust meaning they are much aware of their Jewishness and that was only one category. It would be expected that most of the remaining 27 percent had a similar equally Jewish identity. There are 55 politically oriented Jewish organizations in the United States. [2] And, that's just for politics. Somebody belongs to these groups. The point is, at least in economic activity, most Jews are aware of their Jewishness despite of what the say on surveys.

Jewish Religious Sects

Through the centuries the Jews have, to a remarkable degree, kept themselves as a pure race. The race is one thing, the Jewish religion is another. As an example, President Donald Trump had three children from his first marriage, two boys and one girl. All three married Jews and converted to the Jewish faith. That seems simple enough, but is it? Consider that there are secular Jews, orthodox Jews, observant Jews, and religious Jews. There are Rabbinic Jews, Hasidic Jews, Karaite Jews and The Dönmeh as well as mixtures of these and others. For example, some orthodox Jews are also observant Jews. The secular Jews are those who rarely if ever go to synagogue. However, if a cousin was getting married they'd in all probably show up wearing a yarmulke. The point is that they are all of a race except for a few converts. Some observe the rules of the Torah or the Talmud but have no faith. With only fifteen million Jews in the world it is surprising that as fractionated as they are religiously they manage to stick together so much as a race. Here an effort will be made to untangle some of the terms used to describe Jews and their sects.

Orthodox Versus Religious Jews

Orthodox Jews are often called religious by the media but that is not always accurate. The term "orthodox" implies "observant." However,

---

[1] The demographic data is taken from "Pew Survey of U.S. Jews," *Baltimore Jewish Times*, October 3, 2013, Maayan Jaffe & Uriel Heilman. The data is based on a telephone survey of 3,475 Jews nationwide.

[2] https://www.jewishvirtuallibrary.org/american-jewish-organizations

there is a difference between being observant and being religious. An Orthodox Jew outwardly presents himself as someone who follows the Torah, the written law passed down from Moses at Sinai. Yet, often people who publicly appear Orthodox will act in a manner contrary to religious doctrine, hence, all Orthodox Jews are not religious and the words Orthodox and religious should not be used interchangeably.

Rabbinic Judaism

Rabbinic Jews call themselves "rabaniyin" or "Followers of the Rabbis" or talmudiyin "Followers of the Talmud." That is, this group lets the Rabbis interpret the Talmud for them, and as we shall see later the Talmud is a large body of rules and laws that has a tendency to change over time.

Hasidic Jews

Hasidic (or Chasidic) Judaism is a conservative branch of Haredi Judaism, which is itself a branch of Orthodox Judaism. Thus, Hasidic Jews are Orthodox, although they differ from Orthodox Jews in some respects. The word *Hasidic* comes from the Hebrew word *chesed*, meaning "loving-kindness." The Hasidim are literally "those who do good deeds for others." They are known for their separated living, their devotion to a dynastic leader, their exuberant, joyful worship, and their distinctive dress. Hasidic Jews believe that prayer and acts of loving-kindness are means of reaching God.

Karaite Jews

Karaites are a relatively small sect of Jews who believe in God and look forward to the coming of the Messiah, either bar David or bar Jonah. The word "Karaite" means "Readers of Scripture" and is derived from the old Hebrew word for the Hebrew Bible, Mikra, or Kara. This name was chosen by the adherents of Karaite Judaism to distinguish themselves from the adherents of Rabbinic Judaism.

Ashkenazic Jews

Ashkenazic Jews are the Jews of France, Germany, and Eastern Europe and their descendants. The adjective "Ashkenazic" and corresponding nouns, Ashkenazi (singular) and Ashkenazim (plural) are derived from the Hebrew word "Ashkenaz," which is used to refer to Germany. Most

American Jews today are Ashkenazim, descended from Jews who emigrated from Germany and Eastern Europe from the mid 1800s to the early 1900s.

Sephardic Jews

Sephardic Jews are the Jews of Spain, Portugal, North Africa and the Middle East and their descendants. The adjective "Sephardic" and corresponding nouns Sephardi (singular) and Sephardim (plural) are derived from the Hebrew word "Sepharad," which refers to Spain. Most of the early Jewish settlers of North America were Sephardic.

Mizrachim Jews

Sephardic Jews are often subdivided into Sephardim, from Spain and Portugal, and Mizrachim, from the Northern Africa and the Middle East. The word "Mizrachi" comes from the Hebrew word for Eastern. There is much overlap between the Sephardim and Mizrachim. Until the fifteenth century, the Iberian Peninsula, North Africa and the Middle East were all controlled by Muslims, who generally allowed Jews to move freely throughout the region. When the Jews were expelled from Spain in 1492, many of them were absorbed into existing Mizrachi communities in Northern Africa and the Middle East.

Jews in Israel

In Israel, a little more than half of all Jews are Mizrachim, descended from Jews who have been in the land since ancient times or who were forced out of Arab countries after Israel was founded. Most of the rest are Ashkenazic, descended from Jews who went to the Holy Land (then controlled by the Ottoman Turks) instead of the United States in the late nineteenth century, or from Holocaust survivors, or from other immigrants who came at various times.

Dönmeh Jews

There are also quasi Jewish sects chief among them is The Dönmeh. This sect is an historical monster lurking in the background of almost every serious military and diplomatic incident involving Israel, Turkey, Iran, Saudi Arabia, Iraq, Greece, Armenia, the Kurds, the Assyrians, and other players in the Middle East and southeastern Europe. These Jewish refugees from Spain in the late fifteenth century were welcomed to settle in the Ottoman Empire and over the years they converted to a mystical sect of Islam that eventually mixed Jewish Kabbala and Islamic Sufi semi-mystical beliefs into a sect that eventually championed secularism in post-Ottoman Turkey.

The Dönmeh sect of Judaism was founded in the seventeenth century by Rabbi Sabbatai Zevi who we have discussed earlier. They were crypto-Jews, publicly proclaimed their Islamic faith but secretly practiced their hybrid form of Judaism, which was unrecognized by mainstream Jewish rabbinical authorities. The practice of publicly professing one faith and secretly practicing Judaism is what they were doing in Spain with Christianity and that was the main cause of their expulsion from that country following 1492.

Zionism

It is well to include Zionism here since it is necessary that the term be defined and elaborated upon in this study of Jews. It is a definition, however, that tangentially includes others as well as Jews because it spills over into certain Christian sects. Generally, Zionism includes individuals who believe that the continued existance and prosperity of the State of Israel supercedes all other allegiances. That is, a Zionist puts the good of the State of Israel ahead of his native land. Zionists are primarily Jewish by ethnicity many of whom live in and are citizens of foreign countries. As we shall see further on, there is a disproportionate number of them in the U.S. government, some with dual Israel and U.S. citizenship whose primary mission in life is to use their positions of power to further Israel's interests whether or not it harms America.

Zionism gains its power from being a singular combination of religion, politics and money. Its basic power comes from being a religious misrepresentation that equates the modern Zionist state of Israel with the one found in the Bible. Since Israel's founding in 1948 its leaders have accepted that unlikely event as meaning Israel was to restore the Kingdom of David and Solomon to its biblical borders no matter who was in the way. The money comes mainly from the international banking cabal led by the House of Rothschild. This idea of restoring the "promised land" is believed by millions of well meaning but naïve and misled Christian Zionists most of whom live in the United States.

Chapter 16

The Messiah

The Meaning of Messiah to the Jews

As mentioned at the beginning Part III, at the time of Christ many of the Jews and most of the leaders of the Jews, were expecting a Messiah bar David, a worldly leader, a general, that would free them from the yoke of the Romans. As also mentioned the Old Testament contained prophecies about the coming Messiah with two distinguishable meanings. Some were of the Messiah being a worldly king and leader, that is the Messiah bar David, and the other that of a spiritual leader the Messiah bar Jonah.

When speaking of the Old Testament with respect to the Jews some distinctions in terms are needed. Christians will some times say that to the Jews the Old Testament is the Torah. But, technically Torah refers to only the first five books, the Pentateuch, written by Moses. The entire Old Testament, which for the Jewish is the whole of their scripture, is the Tanakh (pronounced ta-nack') composed of 39 books. Tanakh is an acronym from the Hebrew letters of its three components: Torah (Pentateuch, 5 books), Nevi'im (Prophets, 21 books), and Ketuvim (Writings, 13 books). That is not the end of it, though. The Catholic Old Testament contains 46 books as opposed to the 39 books in the Tanakh. By the way, they are the same 39 books found in the Protestant Bible. The Catholic Bible also contains Tobias, Judith, Wisdom, Sirach (Ecclesiasticus), Baruch, and 1 and 2 Machabees. How the extra seven books came to be in the Catholic Bible is an interesting story of history. If the reader is interested that story can be found in the Appendix.

If a person goes on the Internet and asks, "Are the Jews still expecting a Messiah?" there will be many answers as one would expect but there are some common elements which will be listed here.

The first thing most people do who discuss the topic will be to establish the origin of he word 'messiah' as being not a particularly Jewish concept because they have a different understanding of the Torah, Tanakh and Talmud (much more on the Talmud later). For them messiah is a Westernized form of the Hebrew word 'moshiach' which translates to 'anointed.' The title of moshiach was given to any person who was appropriately anointed with oil as part of their initiation to their service of HaShem, a title used in Judaism to refer to God. The Jews have had many moshiachim in the form of kings, priests, prophets, and judges. And, they hold that there is nothing supernatural about a moshiach. They say that is strictly a Christian connection of messiah with Jesus the Divine. There are prophecies of a future moshiach. However, they hold that it is a minor topic in Judaism and the Tanakh.

They also hold that there are different kinds of "moshiachim." There is Moshiach bar David as mentioned above. There could be several Moshiach bar Yosef to prepare the way for Moshiach bar David. Different iterations of Elijah may come before Moshiach bar David. It is a more complicated tradition than just "yes" or "no" as Christians see it. However, it is an article of Faith that says: I believe in the arrival of moshiach and the era of moshiach. Significantly, they do not mention Moshiach bar Jonah.

The Messiah bar David

As mentioned before, some Jews finally in the nineteenth century began to see the Messiah as the Jewish race taken collectively. For those who still believe in the Messiah as a single man they would list some or all of the following traits he would have.

1. Build the Third Temple.

   I will make with them a covenant of peace; it shall be an everlasting covenant with them, and I will multiply them, and put my sanctuary among them forever. My dwelling shall be with them; I will be their God and they shall be my people. Thus the nations shall know that it is I, the Lord, who make Israel holy, when my sanctuary shall be set up among them forever. Ez. 37: 26-28.

2. Gather all Jews back to the Land of Israel.

Fear not, for I am with you; from the east I will bring back your descendants, from the west I will gather you. I will say to the north: give them up! and to the south: Hold not back! Bring back my sons from afar, and my daughters from the ends of the earth. Is 43: 5-6.

3. Usher in an era of world peace, and end all hatred, oppression, suffering and disease.

He shall judge between the nations and impose terms on many peoples. They shall beat their swords into plowshares and their spears into pruning hooks; one nation shall not raise the sword against another, nor shall they train for war again. Is 2:4.

4. Spread universal knowledge of the God of Israel, which will unite humanity as one.

The Lord shall become king over the whole earth, and that day the Lord shall be the only one, and his name the only one. Zechariah 14:9.

5. Hamoshiach (the anointed) must be descended on his father's side, from King David.

The scepter shall not depart form Judah or the staff from between his feet, until he comes to whom it belongs. To him shall be the obedience of nations. Genesis 49:10.

And there shall come forth a rod out of the root of Jesse, and a flower shall rise up out of his root. Is 11:1.

6. Hamoshiach will lead the Jewish people to full Torah observance. The Torah states that all commands remain binding forever, and anyone coming to change the Torah is immediately identified as a false prophet.

Every command that I enjoin on you, you shall be careful to observe, neither adding to it nor subtracting from it. If there arise

among you a prophet or a dreamer who promises you a sign or
wonder, urging you to follow other gods, whom you have not
known, and to serve them: even though the sign or wonder he has
foretold you comes to pass, pay no attention to the works of that
prophet or that dreamer; for the Lord your God, is testing you to
learn whether you really love him with all your heart and with all
your soul. Duet 13:1-4.

Comments on the Requirement for Messiah

For those of the six conditions as specified above that have only the
biblical references listed, those texts are listed in the comments below.

Condition 1

Concerning the first requirement to build the third temple, if the Jews
are looking for the coming of the Messiah, and he is only an "anointed
one," that has nothing especially divine about him, this requirement will
be impossible. Due to the intransigence of the Jews, the Romans destroyed the temple in AD 70, and dispersed the people. Flavius Claudius
Julianus was the Roman emperor from 361 to 363. He made every effort
to foster religions other than Christianity. Before he died in battle against
the Persians on June 26, 363, Julian appointed his friend and general,
Alypius, to oversee the construction of the third Jewish temple in Jerusalem.

They made everything ready to begin work; workmen were assembled
from many lands; large amounts of materials of various kinds were at
hand. But the bishop, St. Cyril, saw all these preparations without any
concern, relying on the infallible truth of the scripture prophecy "that one
stone should not be left on another." Mt. 24:2.

Cyril predicted with confidence that the Jews, far from being able to
rebuild their ruined temple, would be the very instruments through which
that prophecy of Christ would be more fully accomplish, more than even
the Romans had done, and that they would not be able to put one stone
upon another.

According to the church father Gregory of Nazianzus, writing in
Asia Minor within a year of the project, the Jews "in large number
and with great zeal set about the work.". . .(8) Another contempo-

rary, Ephraem of Syria . . . reported that the Jews "raged and raved and sounded the trumpets" and that "all of them raged madly and were without restraint.(9)

Despite such auspicious beginnings, work on the Temple probably lasted only a few days. Numerous reports, both pagan and Christian, attribute the work stoppage to a fire and, perhaps, an earthquake. The Roman historian Ammianus Marcellinus reported that "terrifying balls of flame kept bursting forth near the foundations of the Temple," burning some of the workers to death and putting a stop to the enterprise. (10) Gregory of Nazianzus wrote of "a furious blast of wind" and "a flame that issued forth from the sacred place." (11) Ephraem noted that there were winds, earthquakes and lightning, and that a "fire came forth.". . . In any event, with Julian's death, the attempt to rebuild the Temple ended. [1]

(8) Gregory of Nazianzus, Oratio V contra Julianum, 4, in C.W. King, trans., Julian the Emperor (London: George Bell and Sons, 1888).

(9) Ephraem of Syria, Hymni contra Julianum, 1.16 and 2.7, in Samuel N.C. Lieu, trans., The Emperor Julian: Panegyric and Polemic (Liverpool, England: Liverpool Univ. Press, 1986).

(10) Ammianus Marcellinus, 23.1,3.

(11) Gregory of Nazianzus, Oratio V contra Julianum, 4.

In view of that, it would reasonably require a miracle to proceed with the building of the third temple. A Messiah with no special supernatural powers could not do that.

Condition 2

This condition lists Isaia 43:5-6 as its authority. If one adds the verse that follows, it can be interpreted to mean everyone on earth, not just the Jews. The entire passage then becomes: "5 Fear not, for I am with you; from the east I will bring back your descendants, from the west I will gather you. 6 I will say to the north: give them up! and to the south: Hold not back! Bring back my sons from afar, and my daughters from the ends

---

[1] http//www.Julian the Apostate and His Plan to Rebuild the Jerusalem Temple, Jeffrey Brodd, BR 11:05, Oct 1995.

of the earth. 7 And every one that calleth upon my name, I have created him for my glory. I have formed him, and made him." Duet 13:1-4.

Conditions 5 and 6

Here the word Hamoshiach is used and is a Hebrew word for God. Moshiach is their word for Messiah which is the title for the liberator of the Jewish people. So, why would Hamoshiach be used here because it implies the Messiah would be divine something the present day Jews flatly deny. This is an example of the difficulty one finds in researching the Jews. There are many sects and many authorities so a definitive answer on any particular subject is difficult.

Another example of a reference to the Messiah bar David in the Old Testament can be found in Ezechiel.

23 And I will set up one shepherd over them, and he shall feed them, even my servant David: he shall feed them, and he shall be their shepherd. 24 And I the Lord will be their God: and my servant David the prince in the midst of them: I the Lord have spoken it. 25 And I will make a covenant of peace with them, and will cause the evil beasts to cease out of the land: and they that dwell in the wilderness shall sleep secure in the forests. 26 And I will make them a blessing round about my hill: and I will send down the rain in its season, there shall be showers of blessing. 27 And the tree of the field shall yield its fruit, and the earth shall yield her increase, and they shall be in their land without fear: and they shall know that I am the Lord, when I shall have broken the bonds of their yoke, and shall have delivered them out of the hand of those that rule over them. 28 And they shall be no more for a spoil to the nations, neither shall the beasts of the earth devour them: but they shall dwell securely without any terror. 29 And I will raise up for them a bud of renown: and they shall be no more consumed with famine in the land, neither shall they bear any more the reproach of the Gentiles. 30 And they shall know that I the Lord their God am with them, and that they are my people the house of Israel: says the Lord God. Ez 34:23-30.

This whole passage especially Verse 29 could easily be interpreted as a reference to a worldly leader: "And I will raise up for them a bud of renown . . . neither shall they bear any more the reproach of the Gentiles." It is also understandable that in hindsight, Christians could see this as a clear reference to Jesus especially in the first part of that verse, "And I will raise up for them a bud of renown. . . ." In my Douay Rheims is the note: He speaks of Christ our Lord, the illustrious bud of the house of David, renowned over all the earth." This is also connected by Christians to Jeremiah, "In those days, and at that time, I will make the bud of justice to spring forth unto David, and he shall be judgement and justice in the earth." Jeremiah 33:15. It says justice in the earth, not only in Israel which is significant as we shall now see.

The Messiah bar Jonah

The Messiah bar Jonah is not mentioned by the Jews much today even though the book of Jonah is in their Tanakh so he is not an invention of Christendom. The important thing here is to go back and try to see what the Jews at the time of Christ thought Messiah meant. There is no difficulty finding that the notion of Messiah bar David was strongly accepted. However, if one looks, another meaning is present, in fact, the whole message of Jesus is other than worldly. For example in Matthew we have:

Now when John had heard in prison the works of Christ: sending two of his disciples he said to him: Are you he that is to come, or look we for another? And Jesus making answer said to them: Go and relate to John what you have heard and seen. The blind see, the lame walk, the lepers are cleansed, the deaf hear, the dead rise again, the poor have the gospel preached to them. And blessed is he that shall not be scandalized in me. And when they went their way, Jesus began to say to the multitudes concerning John: What went you out into the desert to see? A reed shaken with the wind? But what went you out to see? A man clothed in soft garments? Behold they that are clothed in soft garments, are in the houses of kings. But what went you out to see? A prophet? Yea I tell you, and more than a prophet. For this is he of whom it is written: *Behold I send my angel before your face, who shall prepare your way before you. Malachia 3:1.* Amen I say to you, there has not risen among them that are born of women a greater than John the Baptist: yet he that is the lesser in the kingdom of heaven is greater than he. And from the days of John the Baptist until now, the kingdom of heaven suffers violence, and the violent bear it away. For all the prophets and the law prophesied until John: And if you will receive it, he is Elias that is to come. He that has ears to hear, let him hear. Mat. 11:2-14. [2]

The activities of Jesus certainly are not the same as gathering, arming and training an army. So if many were looking for a worldly leader many others were happy to accept the spirituality offered by Jesus. This is brought to a peak in His triumphal entry into Jerusalem the week before the passion where the common people were shouting His praises in a public display of support.

And they brought the ass and the colt and laid their garments upon them and made him sit thereon. And a very great multitude spread their garments in the way: and others cut boughs from the trees and strewed them in the way. And the multitudes that went before and that followed cried, saying: *Hosanna to the son of David: Blessed is he that comes in the name of the Lord: Hosanna in the highest, Ps 117:26f.* And when he had come into Jerusalem, the whole city was moved, saying: Who is this? And the people said: This is Jesus, the prophet from Nazareth of Galilee. Mat. 21: 7-11.

The title Son of David is clearly Messianic. This is the first time Jesus permits Himself to be publicly seen and proclaimed as the Messiah. Hosanna addressed to God means "Save us, we pray." All four Gospels record this event. It must be borne in mind that the title Son of David as used here is not to be seen as David vs. Jonah as we have been discussing, but strictly as a Messianic designation.

These and many more references could be cited that make it clear that one was awaited who was to be something other then a general, a

---

[2] Nearly identical words are used in Luke 7: 20-28.

worldly leader. The Messiah bar Jonah is most clearly seen, of course, in the book of Jonah from which comes the appellation. Here Jonah was sent by God to the city of Nineveh the capital city of the Assyrian Empire, the traditional enemy of Israel. It was located on the west bank of the Tigris River in the northern part of modern day Iraq. It was a large city, "Now Nineveh was an enormously large city; it took three days to go through it." Jonah 3:3. The population of the city was large for ancient times. ". . . in which there are more than 120,000 persons." Jonah 4:11. The narrative shows Nineveh as a wicked city worthy of destruction. God sent Jonah to preach to them of their coming destruction; they fasted and repented because of this. As a result, God stayed His wrath. At this Jonah, a Jew, remonstrated with God as in this passage.

But this was greatly displeasing to Jonah, and he became angry. "I beseech you, Lord," he prayed, "is not this what I said while I was still in my own country? This is why I fled at first to Tharsis. I knew that you are a gracious and merciful God, slow to anger, and rich in clemency, loathe to punish. And now, Lord, please take my life from me, for it is better for me to die than to live." But the Lord asked, "Have you reason to be angry?" Jonah 4:1-3.

God is stating that He is showing mercy for the population who are ignorant of the difference between right and wrong (who cannot distinguish their right hand and their left) and even showed mercy for the animals in the city.

What was the real reason why Jonah was angry with God? It was because He did something good for Gentiles. Jonah had taken to heart that Yahweh set up a covenant where He was the God of the Israelites and they were His people. The Israelites took this to mean they had an exclusive relationship with God, but here God is saying that, yes, He is the God of the Israelites, but He is also the God of everyone else. That was Jonah's complaint. The significance of this book is that the Assyrians were, fist of all, not Hebrews and, secondly, were not in danger of being defeated by enemies or in any other physical danger. They were in moral danger. As such, it prepared the way for the Gospel with its message of redemption for all, both Jew and Gentile. Interestingly, Nineveh's repentance and salvation can be found not only in the Christian Bible, but also the Jewish Tanakh as we have seen, and the Muslim Quran. Yet, the Jews do not consider Gentiles as being redeemable, and the Muslims feel the same about Infidels. Was the rejection of Jesus by the Jews perhaps as much about the fact that the Messiah bar Jonah was a savior of everyone as it was that He was not a worldly leader?

The Passion of the Christ

There were also prophecies of how the Messiah bar Jonah had to suffer and die for his people. After Jesus' public life of teaching, drawing crowds to himself, causing almost mass conversions, and working countless miracles his passion is the final proof that He was indeed the long awaited Messiah.

Earlier we saw that at the trial of Jesus before Pilot the two notions of the Messiah were juxtaposed when Pilot offered Barabbas as an alternative to Jesus.

Pilate, indeed, did not fail to make attempts to liberate Christ, but they were without result. Neither his eloquence in repeatedly defending the innocence of Christ, nor the shrewd scheme of opposing Jesus to Barabbas, nor the resort to scourging, nor the Ecce Homo had availed anything. Now he took the last refuge in sneers and ridicule. The Jews had said, "If thou release this man, thou are not Caesar's friend. For whosoever maketh himself a king, speaketh against Caesar." Pilate had somewhat recovered from the terror which these words had injected into him and slyly pretended not to have understood anything of their threats. He now railed at the foolishness of the Jews in considering as their king such a piteously mangled being. "Behold your King," said he, "he, indeed looks like a king!" Carried away by anger at this insult, not the high-priests alone, but all present cried together, "Away with him, away with him, crucify him."[3]

One has to, in a way, feel sorry for the Chief Priests. On the one hand Jesus went out of his way to rail against them frequently calling them

---

[3] James Groenings, S.J. *The History Of The Passion*, South Bend, Indiana, Marian Publications, 1908, p. 233.

hypocrites, tripping them up in their speech when they tried to trap Him and ultimately throwing the money changers out of the temple. Yet, they had to be aware of the fact that He was no ordinary man. If He was not God then He was the most exalted man ever created. Some of their number were on hand to witness miracles and otherwise they had heard of Him curing lepers, and even bring to life someone, Lazarus, who had been dead for days and was decomposing. These actions went beyond what a magician could do. Add to that the huge following he had because of his message. He was preaching that the humble were the ones loved by God and would get to heaven, not the ostentatious Chief Priests, Elders, Scribes, and Pharisees. They were human beings so they had to have been torn. At bottom when the Messiah came, He did not fit their idea of what the Messiah was to be, that is, a worldly leader—for them.

It is a worthwhile meditation to reflect on the scene of the crucifixion vis-à-vis the Chief Priests, Elders, Scribes, and Pharisees. They are standing by jeering at him, "He saved others, let Him save Himself." Mat. 27:42. [4] At the same time they look on in horror as the soldiers go about dividing his garments, and casting lots for his vesture. The rattle of the dice must have stung them deeply. The words of Psalm 21, would have instantly come to mind:

> For many dogs surround me:
> a pack of evildoers closes in upon me.
> They have pierced my hands and feet.
> They have numbered all my bones.
> And they look on and gloat over me.
> They divide my garments amongst them;
> and upon my vesture they cast lots.
>   Ps. 21:17-19.

What was happening? Were they, in fact, the "dogs and evildoers" predicted? All through the past night and this day they had one driving desire and that was to be rid of Jesus. Now nearing the conclusion of their endless deceit and lying was it all to be undone? The unprecedented darkness that fell on the earth tore at their hearts as if it were a premoni-

---

[4] Nearly identical language in Mk. 15:30; Lk. 24:23.

tion of their deaths and the possibility of their hell. They stood their posts, though, to ensure His disciples did not intervene. Then after hours of silence Jesus speaks in a loud voice, "My God, my God, why have you forsaken me?" Ps. 12:2. in a clear reference to the same Psalm. There was no mistaking it now, He was referring to them; that Psalm was written specifically to describe them and it was being fulfilled in their hearing.

Their shallow attempt to deflect the meaning of what Christ said by pretending that he was calling on Elias must have drawn snickers from some of the onlookers. When Jesus said, *Eli, Eli, lema sabacthani*, no Jew hearing this would have though he was saying Elias, both from the standpoint that the two words were so different, and because it made no sense for him to be saying "Elias, Elias, why have you forsaken me." Elias was certainly a prophet of note, but it was not the custom for Jews to pray to him in the same way they prayed to God. How is it that they could have become so blinded by their desires that all shred of common sense left them? Has that at times happened, at least partially, to every one of us?

It is well to expand on the point that Jesus, Messiah bar Jonah, was not only a spiritual leader but He had come to save all of mankind. The Jewish leaders knew that was the thrust of his teaching, and He backed up his words with miracles, something they could not deny. They had sent representatives to keep track of this phenomena called Jesus. They obviously spent time thinking up trick questions to trap him as in their question about whether it was lawful to give tribute to Caesar. Mt. 22:17. [5] He cured the servant of the centurion, who was not a Jew. Mt. 8:5-13; [6] Even in the sermon on the mount when he said, "Blessed are they who mourn, for they shall be comforted." Mt. 5:5. He clearly did not say, "Blessed are the Jews who mourn . . . ." Yet, not only is the distinction of the Messiah bar David versus Bar Jonah not taught today, the clear change from the Old Testament "Chosen people" to the choosing of all people is given little attention. This has led to the heresy prevalent in the Catholic Church that the Jews are even today being given special treatment when it comes to salvation.

---

[5] This event is also recorded in Mk. 12:14; Lk. 20:22.
[6] Also Lk. 7:1-10.

With that in mind the twenty-first Psalm seemed to have been written for that moment on Calvary. Rarely was a prophecy fulfilled in such literal detail. That being the case verses twenty-eight through thirty once again bring out the case where the Messiah bar Jonah is not only a spiritual leader and savior, but he comes for the whole of mankind, not exclusively for the Israelites.

> All the ends of the earth shall remember,
> and shall be converted to the Lord:
> And all the kindreds of the Gentiles shall
> bow down before him.
> For the kingdom is the Lord's;
> and he shall have dominion over the nations.
> To him alone shall bow down
> all who sleep in the earth;
> Before him shall bend all who go down into the dust.
> Ps 21:28-30.

Chapter 17

Their Writings

The Talmud

To understand the Jews it is necessary to understand how they think and what drives their beliefs. As it turns out there is a book that is the basis for their thought just as the Bible is the basis for the Christian religions. Few people would know what you were talking about if you mentioned the Talmud. Those a little in the know might say, "You mean the Torah?" You would have to explain that the two are quite different writings. What follows are some excerpts from a piece put on the Internet by Michael A. Hoffman II and Alan R. Critchley. It is representative of other sources but the two authors have collected many of the significant points in a small space. What follows are selected passages from that web site to provide an idea of what is in the Talmud. The reader is encouraged to investigate further by visiting this and similar web sites. [1] The authors' references are left to stand by themselves.

Michael A. Hoffman II and Alan R. Critchley, Copyright ©2000. All Rights Reserved. Independent History & Research, Box 849, Coeur d'Alene, Idaho 83816. http://www.hoffman-info.com/talmudtruth.html.

---

[1] http://www.catholicapologetics.info/apologetics/judism/ There are several subsections at this site dealing with the Talmud that say substantially the same thing as Hoffman and Critchley.

Introduction

The Talmud is Judaism's holiest book (actually a collection of books). Its authority takes precedence over the Old Testament in Judaism. Evidence of this may be found in the Talmud itself, Erubin 21b (Soncino edition): "My son, be more careful in the observance of the words of the Scribes than in the words of the Torah (Pentateuch)."

Rabbi Joseph D. Soloveitchik is regarded as one of the most influential rabbis of the 20th century, the "unchallenged leader" of Orthodox Judaism and the top international authority on *halakha* (Jewish religious law). Soloveitchik was responsible for instructing and ordaining more than 2,000 rabbis, "an entire generation" of Jewish leadership.

N.Y. Times religion reporter Ari Goldman described the basis of the rabbi's authority: "Soloveitchik came from a long line of distinguished Talmudic scholars. . . . He came to Yeshiva University's Elchanan Theological Seminary where he remained the pre-eminent teacher in the Talmud. . . .He held the title of Leib Merkin professor of Talmud." (*N.Y. Times*, April 10, 1993, p. 38).

Nowhere does Goldman refer to Soloveitchik's knowledge of the Bible as the basis for being one of the leading authorities on Jewish law. The rabbi's credentials are all predicated upon his mastery of the Talmud. Other studies are clearly secondary."

Having established that it is not hard to find reparable sources that say the Talmud is the primary source of Jewish religions thought today Messieurs Hoffman and Critchley continue with a discussion of how the Bible and the Talmud differ.

The Talmud Nullifies the Bible

The Jewish Scribes claim the Talmud is partly a collection of traditions Moses gave them in oral form. These had not yet been written down in Jesus' time. Christ condemned the traditions of the Mishnah (early Talmud) and those who taught it (Scribes and Pharisees), because the Talmud nullifies the teachings of the Holy Bible.

The famous warning of Jesus Christ about the tradition of men that voids Scripture (Mark 7:1-13), is in fact, a direct reference to the Talmud, or more specifically, the forerunner of the first part of it, the Mishnah.

For those who rarely interrupt their reading to look up a reference, even a readily available one like a biblical text, the reference above will be reproduced here so the reader can see Our Lord's pointed condemnation of the Mishnah.

1 And there assembled together unto him the Pharisees and some of the scribes, coming from Jerusalem. 2 And when they had seen some of his disciples eat bread with common, that is, with unwashed hands, they found fault. 3 For the Pharisees and all the Jews eat not without often washing their hands, holding the tradition of the ancients. 4 And when they come from the market, unless they be washed, they eat not: and many other things there are that have been delivered to them to observe, the washings of cups and of pots and of brazen vessels and of beds. 5 And the Pharisees and scribes asked him: Why do not your disciples walk according to the tradition of the ancients, but they eat bread with common hands? 6 But he answering, said to them: Well did Isaia prophesy of you hypocrites, as it is written: *This people honors me with their lips, but their heart is far from me. 7 And in vain do they worship me, teaching doctrines and precepts of men.* 8 For leaving the commandment of God, you hold the tradition of men, the washing of pots and of cups: and many other things you do like to these. 9 And he said to them: Well do you make void the commandment of God, that you may keep your own tradition. 10 For Moses said: *Honor your father and your mother.* And *He that shall curse father or mother, dying let him die.* 11 But you say: If a man shall say to his father or mother, Corban (that is given to God) whatsoever is from me shall profit you. 12 And further you allow him not to do anything for his father or mother, 13 making void the word of God by your own tradition, which you have given forth. And many other such like things you do. Mark 7:1-13

Messieurs Hoffman and Critchley continue:

Unfortunately, due to the abysmal ignorance of our day, the widespread "Judeo-Christian" notion is that the Old Testament is the supreme book of Judaism. But this is not so. The Pharisees teach for doctrine the commandments of rabbis, not God.

The Talmudic commentary on the Bible is their supreme law, and not the Bible itself. That commentary does indeed, as Jesus said, void the laws of God, not uphold them. As students of the Talmud, we know this to be true. . . .

To the Mishnah the rabbis later added the Gemara (rabbinical commentaries). Together these comprise the Talmud. There are two versions, the Jerusalem Talmud and the Babylonian Talmud.

The Babylonian Talmud is regarded as the authoritative version: "The authority of the Babylonian Talmud is also greater than that of the Jerusalem Talmud. In cases of doubt the former is decisive." (R. C. Musaph-Andriesse, *From Torah to Kabbalah: A Basic Introduction to the Writings of Judaism*, p. 40).

This study is based on the Jewish-authorized Babylonian Talmud. We have published herein the authenticated sayings of the Jewish Talmud. Look them up for yourself.

Hoffman and Critchley proceed with a long list of specific cases of what the Talmud means to Jews and the rest of human society. A few such instances will be included below as illustration.

Jesus in the Talmud

While it is the standard disinformation practice of apologists for the Talmud to deny that it contains any scurrilous references to Jesus Christ, certain Orthodox Jewish organizations are more forthcoming and admit that the Talmud not only mentions Jesus but disparages him (as a sorcerer and a demented sex freak). These orthodox Jewish organizations make this admission perhaps out of the belief that Jewish supremacy is so well-established in the modern world that they need not concern themselves with adverse reactions.

For example, on the website of the Orthodox Jewish Hasidic Lubavitch group—one of the largest in the world—we find the following statement, complete with Talmudic citations:

The Talmud (Babylonian edition) records other sins of 'Jesus the Nazarene':
1) He and his disciples practiced sorcery and black magic, led Jews astray into idolatry, and were sponsored by foreign, gentile powers for the purpose of subverting Jewish worship (Sanhedrin 43a).
2) He was sexually immoral, worshipped statues of stone (a brick is mentioned), was cut off from the Jewish people for his wickedness, and refused to repent (Sanhedrin 107b; Sotah 47a).

Only two of many supposed sins of Jesus are included here as examples of the calumny that the Talmudic Jews dump on their chief adversary. Hoffman and Critchley continue.

Talmud Attacks Christians and Christian Books
Rosh Hashanah 17a. Christians (*minnim*) and others who reject the Talmud will go to hell and be punished there for all generations.

Sanhedrin 90a. Those who read the New Testament ("uncanonical books") will have no portion in the world to come.

Shabbath 116a. Jews must destroy the books of the Christians, i.e. the New Testament.

Genocide Advocated by the Talmud
Minor Tractates. Soferim 15, Rule 10. This is the saying of Rabbi Simon ben Yohai: *Tob shebe goyim harog* ("Even the best of the gentiles should all be killed"). This passage is from the original Hebrew of the Babylonian Talmud as quoted by the 1907 *Jewish Encyclopedia,* published by Funk and Wagnalls and compiled by Isidore Singer, under the entry, "Gentile," (p. 617). . . .

Israelis annually take part in a national pilgrimage to the grave of Simon ben Yohai, to honor this rabbi who advocated the extermination of non-Jews. (*Jewish Press*, June 9, 1989, p. 56B).

On Purim, Feb. 25, 1994, Israeli army officer Baruch Goldstein, an orthodox Jew from Brooklyn, massacred 40 Palestinian civilians, including children, while they knelt in prayer in a mosque.

Goldstein was a disciple of the late Brooklyn Rabbi Meir Kahane, who told CBS News that his teaching that Arabs are "dogs" is derived "from the Talmud." (CBS *60 Minutes*, "Kahane").

University of Jerusalem Prof. Ehud Sprinzak described Kahane and Goldstein's philosophy: "They believe it's God's will that they commit violence against *goyim*, a Hebrew term for non-Jews." (*NY Daily News*, Feb. 26, 1994, p. 5).

Talmudic Doctrine: Non-Jews are not Human

The Talmud specifically defines all who are not Jews as non-human animals, and specifically dehumanizes Gentiles as not being descendants of Adam. Here are some of the Talmud passages which relate to this topic.

Kerithoth 6b: Uses of Oil of Anointing. "Our Rabbis have taught: He who pours the oil of anointing over cattle or vessels is not guilty; if over gentiles (goyim) or the dead, he is not guilty. The law relating to cattle and vessels is right, for it is written: "Upon the flesh of man (Adam), shall it not be poured (Exodus 30:32); and cattle and vessels are not man (Adam). [Exodus 30:32 is as follows: The flesh of man shall not be anointed therewith, and you shall make none other of the same composition, because it is sanctified, and shall be holy unto you.]

"Also with regard to the dead, (it is plausible) that he is exempt, since after death one is called corpse and not a man (Adam). But why is one exempt in the case of gentiles (goyim); are they not in the category of man (Adam)? No, it is written: 'And ye my sheep, the sheep of my pasture, are man" (Adam); [Ez 34:31]: Ye are called man (Adam) but gentiles (goyim) are not called man (Adam)."

In the preceding passage, the rabbis are discussing the portion of the Mosaic law which forbids applying the holy oil to men.

Ezechiel 34:31 is the alleged Biblical proof text repeatedly cited in the preceding Talmud passages. But Ezechiel 34:31 does not in fact support the Talmudic notion that only Israelites are human.

The actual text from Ezechiel 34:31 from the Douay Rheims Bible is as follows: "And you my flocks, the flocks of my pasture are men: and I am the Lord your God, saith the Lord God." In The New American Bible, St. Joseph edition, it is rendered this way: "You, my sheep, you are the sheep of my pasture, and I am your God, says the Lord God," The significance of the omission of the two words "are men" in the American Bible is open to question. Notes that accompany my edition of the Douay Rheims states that in this case "men" only signifies human beings, that is, all human beings. It might be as simple as when Ezechiel, under the inspiration of God, wrote that verse with the references to flocks he wanted to make it certain that he was, in fact, talking about people, not sheep. Now to continue with Messieurs Hoffman and Critchley.

Moses Maimonides: Advocate of Extermination

We will now examine the post-Talmudic commentator Rambam (Moses Maimonides). This revered "sage" taught that Christians should be exterminated.

Moses Maimonides is considered the greatest codifier and philosopher in Jewish history. He is often affectionately referred to as the Rambam, after the initials of his name and title, Rabenu Moshe Ben Maimon, "Our Rabbi, Moses son of Maimon." (*Maimonides' Principles*, edited by Aryeh Kaplan, Union of Orthodox Jewish Congregations of America, p. 3). According to *Maimonides' Principles,* p. 5, Maimonides spent twelve years extracting every decision and law from the Talmud, and arranging them all into 14 systematic volumes. The work was finally completed in 1180, and was called Mishnah Torah, or "Code of the Torah."

Here is what Maimonides (Rambam) taught concerning saving people's lives, especially concerning saving the lives of gentiles and Christians, or even Jews who dared to deny the "divine inspiration" of the Talmud:

Maimonides, *Mishnah Torah,* (Moznaim Publishing Corporation, Brooklyn, New York, 1990, Chapter 10, English Translation), p. 184: "Accordingly, if we see an idolater (gentile) being swept away or drowning in the river, we should not help him. If we see that his life is in danger, we should not save him." The Hebrew text of the Feldheim 1981 edition of *Mishnah Torah* states this as well.

Immediately after Maimonides' admonition that it is a duty for

Jews not to save a drowning or perishing gentile, he informs us of the Talmudic duty of Jews towards Christians, and also towards Jews who deny the Talmud. Maimonides, *Mishnah Torah*, (Chapter 10), p. 184: "It is a *mitzvah* (religious duty), however, to eradicate Jewish traitors, *minnim*, and *apikorsim*, and to cause them to descend to the pit of destruction, since they cause difficulty to the Jews and sway the people away from God, as did Jesus of Nazareth and his students, and Tzadok, Baithos, and their students. May the name of the wicked rot."

Jewish Deception and Dissimulation

The response of the orthodox rabbis to documentation regarding the racism and hatred in their sacred texts is simply to brazenly lie, in keeping with the Talmud's Baba Kamma 113a which states that Jews may use lies ("subterfuge") to circumvent a Gentile.

In 1994, Rabbi Tzvi Marx, director of Applied Education at the Shalom Hartman Institute in Jerusalem, made a remarkable admission concerning how Jewish rabbis in the past have issued two sets of texts: the authentic Talmudic texts with which they instruct their own youth in the Talmud schools (yeshiviot) and "censured and amended" versions which they disseminate to gullible non-Jews for public consumption.

Rabbi Marx states that in the version of Maimonides' teachings published for public consumption, Maimonides is made to say that whoever kills a human being transgresses the law.

But, Rabbi Marx points out "...this only reflects the censured and amended printed text, whereas the original manuscripts have it only as 'whoever kills an Israelite."(*Tikkun: A Bi-Monthly Jewish Critique* May-June, 1994).

"Judeo-Christian" Response to the Talmud

Neither the modern popes or the modern heads of Protestantism, have ever insisted that the rabbis of Judaism repudiate or condemn the racism in the Talmud or the murderous hate for Christians and gentiles expressed within it. On the contrary, the heads of Christianity have urged the followers of Christ to obey, honor and support the followers of the Talmud.

This rather long citation about the Talmud has been included to show not only the content, but the many sources of the material, many of which are secular. This thinking is hardly a secret. One must ask how many Jews take the sentiments expressed seriously. It would seen certain that nearly all would have heard about the Talmud, even the two-thirds who, in the survey described in the section on demographics, said they are not raising their children Jewish at all. Even if they are not raising their children Jewish it is certain that Jews have a strong tendency to help one another economically and professionally. And, the includes nearly all of them. It is beyond question that Jews are substantially over represented in all professional fields including university professors, law, medicine, judges, banking, and government to name a few. Is that an accident or do they really have an agenda? It is nice to have a good job, or more correctly, a profession. But, why is their presence in these areas so out of proportion to their numbers in the general population? Make no mistake, every Jew knows he is a Jew, and that is to be expected. After all, everyone of Irish extraction is aware of his lineage. That being so, they find their kind and it should come as no surprise that they help one another and the sentiments expressed in the Talmud are always there.

The Talmud considers the Jewish race superior to the Gentiles. A Jewish author, writing on the influence of the Talmud, has noted, "The men of the other nations are in His eyes on a plane inferior to the Hebrews. It is only by a concession that they have a share in the divine munificence, since only the souls of the Jews descend from the first man. The possessions that are assigned to the other nations in reality belong to Israel." [2] It's just that the Jews haven't yet gotten around to collecting what other nations possess.

It the reader thinks there is a bit of hubris in that last statement consider the following.

"You Will Own Nothing and Be Happy," said Klaus Schwab, Founder and Executive Chairman of the World Economic Forum.

---

[2] Denis Fahey, *The Kingship of Christ and The Conversion of the Jewish Nation* http://www.catholicapologetics.info/apologetics/judism/conversion.htm#VII. Fahey is referring to *L'Harmonie entre l'Eglise et la Synagogue,* by the ex-Rabbin Drach, Vol. I, pp. 9, 223.

As Anthony P. Mueller, a professor of economics, warns, "The main thrust of the forum is global control. Free markets and individual choice do not stand as the top values, but state interventionism and collectivism. Individual liberty and private property are to disappear from this planet by 2030."

America is standing in the way. Critical Race Theory will make sure everyone gets the same piece (equity) of nothing. Communities will be devastated again as jobs are shipped overseas, and foreclosures prevail. People will be pushed to rent. Neighborhoods will be bought up, demolished and high rises will be built. Single-Family Zoning will be outlawed. This will be the last transfer of wealth as the middle class by design once again loses their life savings and private property. Americans will be deemed to be domestic terrorists with racist intentions. Social Emotional Learning has created a class of paid and volunteer activists (rioters). Everyone knows what to do. They learned it in school.[3]

The preceding was said with regard to what Schwab and his fellow one-worlders at the WEF predict they will accomplish when they force Agenda 2030 on the world. The World Economic Forum is composed of the super-rich, mostly Jews, who are through and though communists regardless of what they say. Remember that under communism the individual owns nothing since all property is owned by "everybody." It has never worked that way and it never will. The ultra-rich will own everything and dispose of it as the whim strikes them. It is slavery by a different name. Much more on communism in the following chapter.

The Kabbalah

In the above reference is made to the Kabbalah. Kabbalah means "what has been handed down." Kabbalah is the name applied to the whole range of Jewish mystical writings. While codes of Jewish law focus on what it is God wants from man, kabbalah tries to penetrate deeper, to God's essence itself. In that respect, kabbalah represents an addendum of the Talmud. There are elements of kabbalah in the Bible, for example, in the first chapter of Ezechiel, where the prophet describes his experience of the divine: "...the heavens opened and I saw a divine vision.... As

---

[3] https://www.americaoutloud.com/you-will-own-nothing-and-be-happy-klaus-schwab/

I looked a storm wind came from of the north, a huge cloud of flashing fire, from the midst of which something gleamed like electrum. Within it were figures resembling four living creatures." Ez 1:1,4. The prophet then goes on to describe the creatures, a chariot and the throne of God.

The rabbis of the Talmud regarded the mystical study of God as something that should be done yet was dangerous. There were stories of individuals who became mentally unbalanced while engaging in these mystical activities. The disaster of the false Messiah, Sabbatai Zevi, as mentioned in the short history above, caused seventeenth-century rabbis to legislate that kabbalah should be studied only by married men over forty who were also scholars of Torah and Talmud.

The most famous work of kabbalah, the *Zohar*, was introduced to the Jewish world in the thirteenth century by Moses De Leon, who said the book contained the mystical writings of the second-century rabbi Simeon ben Yochai. Yet many Orthodox mystics see ben Yochai as the recorder of mystical traditions dating back to the time of Moses.

The *Zohar* is written in Aramaic in the form of a commentary on the books of the Torah. The Torah can be seen as a narrative and legal work, but mystics will frequently interpret it as a symbolic representation of the secret laws of the universe and even the secrets of God. This is seen in the way medieval kabbalists, spoke of God as *That Which Is Without Limit*. They predicated that God reveals Himself to us through a series of ten emanations. The first of these is *crown* and refers to God's will to create. Another, *understanding*, represents the unfolding in God's mind of the details of creation. Most of these emanations are regarded as legitimate objects for human meditation through which humans can bring divine grace to the world.

Kabbalah has been one of the important areas of Jewish thought. It contains notions that Jews of today might think of as non-Jewish ideas. One of these is the belief in reincarnation. Between about the sixteenth and nineteenth centuries, it was widely considered by many to be the true Jewish theology. The advent of the modern world in which rational thinking was more highly regarded than the mystical, kabbalah tended to be ignored. However in the last years there has been a renewed interest in the kabbalah, and today it is regularly studied by some non-Orthodox Jews including Hsdidic Jews.

Chapter 18

Communism

We have previously discussed communism in various ways particularly as the Jews brought it to fulfillment in the Judeo-Bolshevik revolution in Russia in 1917. Before we go further we should be certain of what the word communism means and a related word, socialism.

Communism versus Socialism

Today in America one cannot mention communism without being looked upon as retarded because everyone *knows* that communism died when the USSR fell apart. However, socialism is a politically correct word that leftist politicians frequently use to describe even themselves. What, then, is the difference between the two words? In the dictionary [1] one finds this. *Communism* - 1. any economic theory or system based on the ownership of all property by the community as a whole. 2. a hypothetical state of socialism to be characterized by a classless and stateless society and the equal distribution of economic goods and to be achieved by revolutionary and dictatorial, rather than gradualisitc means. *Socialism* - 1. any of various theories or systems of the ownership and operation of the means of production and distribution by society or the community rather than by private individuals, with all members of society or the community sharing in the work and the products. 2. the stage of society, in Marxist doctrine, coming between the capitalist stage and the

---

[1] *Webster's New World Dictionary*, Second College Edition, New York, NY, The World Publishing Company, 1970.

communist stage in which private ownership of the means of production and distribution has been eliminated.

So, in one sense socialism is communism not quite achieved. In another sense they are identical in that both mean there is no private ownership of the means of the production and distribution of goods, but at least at the beginning socialism might allow the private ownership of other property.

Beyond those definitions communism is the form of socialism that uses class warfare and economics to bring about revolution and is international in scope as in the communism in the USSR. Socialism is more of a nationalized version of the same thing that empathizes race as was done in the National Socialists of Germany in the period 1933-45. It is important to understand that in practice both communism and socialism there is total tyrannical rule by a self appointed elite.

Pope Leo XIII addresses socialism in his encyclical on capital and labor. Some passages are relevant here.

> 3. . . . .the ancient workingmen's guilds were abolished in the last century, and no other protective organization took their place. Public institutions and the laws have set aside the ancient religion. Hence, by degrees it has come to pass that working men have been surrendered, isolated and helpless, to the hard heartedness of employers and the greed of unchecked competition. The mischief has been increased by rapacious usury, which, although more than once condemned by the Church, is nevertheless, under a different guise . . . is still practiced by covetous and grasping men . . . .

> 4. To remedy these wrongs the socialists, working on the poor man's envy of the rich, are striving to do away with private property, and contend that individual possessions should become the common property of all, to be administered by the State or by municipal bodies. They hold that by thus transferring property from private individuals to the community, the present mischievous state of things will be set to rights, inasmuch as each citizen will then get his fair share of whatever there is to enjoy. But their contentions are so clearly powerless to end the controversy that were they carried into effect the working man himself would be among the first to suffer. . . .

5. It is surely undeniable that, when a man engages in remunerative labor, the impelling reason and motive of his work is to obtain property, and thereafter to hold it as his very own. If one man hires out to another his strength or skill, he does so for the purpose of receiving in return what is necessary for the satisfaction of his needs; he therefore expressly intends to acquire a right full and real, not only to the remuneration, but also to the disposal of such remuneration, just as he pleases. Thus, if he lives sparingly, saves money, and, for greater security, invests his savings in land, the land, in such case, is only his wages under another form; and, consequently, a working man's little estate thus purchased should be as completely at his full disposal as are the wages he receives for his labor . . . . Socialists, therefore, by endeavoring to transfer the possessions of individuals to the community at large, strike at the interests of every wage-earner, since they would deprive him of the liberty of disposing of his wages, and thereby of all hope and possibility of increasing his resources and of bettering his condition in life. [2]

Here it is seen that all private property is, in a sense, a means of production. It does not have to be land or a factory. Even a keepsake, once owned, could at a later time be sold to generate income.

The Communist Manifesto

The Communist Manifesto published by Karl Marx and Frederick Engels in 1848 is one of those publications like Darwin's *The Origin of Species by Natural Selection.* That is, nearly everyone has heard of it but few have actually read it. However, unless one is engaged in political theory or history, it makes rather boring reading. It is not a long tract, twenty to twenty-five pages depending on the page and type size. It was to be an initial blueprint of what a purely proletarian society would look like. However, today it is most known for the ten measures needed to reform production to the needs of the proletariat. In section II. PROLETARIANS AND COMMUNISTS we find this:

---

[2] Pope Leo XIII, *Rerum Novarum* May 15, 1891.

These measures will of course be different in different countries. Nevertheless in most advanced countries, the following will be pretty generally applicable.

1. Abolition of property in land and application of all rents of land to public purposes.

2. A heavy progressive or graduated income tax.

3. Abolition of all rights of inheritance.

4. Confiscation of the property of all emigrants and rebels.

5. Centralization of credit in the hands of the State, by means of a national bank with State capital and an exclusive monopoly.

6. Centralization of the means of communication and transport in the hands of the State.

7. Extension of factories and instruments of production owned by the State: the bringing into cultivation of waste lands, and the improvement of the soil generally in accordance with a common plan.

8. Equal obligation of all to labor. Establishment of industrial armies, especially for agriculture.

9. Combination of agriculture with manufacturing industries: gradual abolition of all the distinction between town and country, by a more equable distribution of the population over the country.

10. Free education for all children in public schools. Abolition of children's factory labor in its present form. Combination of education with industrial production, &c., &c.

A note should be made here that the ten measures given above were not original with Marx. They appear in a somewhat different form in *The Illuminati Manifesto* and other writings by Adam Weishaupt in the late eighteenth century. Weishaupt was taken up with the French Revolutionary spirit where these ideas flourished. Also note that Marx was a member of the Illuminati whose goal is to have an elite few rule the world. In 1933 the insignia of the Illuminati, the symbol with the all seeing eye at the top of the pyramid, was placed on the back of the American one dollar bill. At the bottom are the words *Novus Ordo Seclorum* which explains the nature of the Illuminati: a "New Social Order" or a "New World Order" which is where we are on the brink of arriving.

The publisher of the November 2006 online version of *The Communist Manifesto* adds comments some of which follow.

The purpose of the measures, dictated at the time by capitalism's unfinished job of "entirely revolutionizing the mode of production," was "to wrest, by degrees, all capital" from the capitalists. Later, when Marx and Engels wrote their joint preface to the German edition of 1872 (quoted by Engels in his preface to the English edition), they expressly stated that this section of the Manifesto would be "very differently worded today." The gigantic strides of industry, the growth of working-class organization, and the practical experience gained in the February Revolution of 1848 and the Paris Commune of 1871, antiquated the ten points. The Marx-Engels revolutionary stand, as opposed to reformism, was certified when the founders of scientific Socialism wrote: "One thing especially was proved by the Commune, viz., that 'the working class cannot simply lay hold of the ready-made State machinery, and wield it for its own purposes.'" [3]

In spite of the above comments in 2006 it is, at times, predicated that the United States is now economically mostly a communist country and on the way to becoming one politically with all the despotism that entails. Following are the measures of *The Communist Manifesto* and the U.S. statutes that would seem to instill them into law. Lists vary but these give the general idea.

1. *Abolition of private property and the application of all rent to public purpose.*
The 14th Amendment of the U.S. Constitution (1868), and various zoning, school & property taxes. [To have a society at all it is necessary to collect some taxes to pay for public works.]
2. *A heavy progressive or graduated income tax.*
The 16th Amendment of the U.S. Constitution, 1913, The Social Security Act of 1936.; Joint House Resolution 192 of 1933; and

---

[3] Hhttp://www.slp.org, Karl Marx and Frederick Engels, Communist Manifesto, Published Online by Socialist Labor Party of America, November 2006.

various State "income" taxes.

3. *Abolition of all rights of inheritance.*
We call it Federal & State estate Tax (1916); or reformed Probate Laws, and limited inheritance via arbitrary inheritance tax statutes.

4. *Confiscation of the property of all emigrants and rebels*

We call it government seizures, tax liens, Public "law" 99-570 (1986); Executive order 11490, sections 1205, 2002 which gives private land to the Department of Urban Development; the imprisonment of "terrorists" and those who speak out or write against the "government" (1997 Crime/Terrorist Bill); or the IRS confiscation of property without due process.

5. *Centralization of credit in the hands of the State, by means of a national bank with state capital and an exclusive monopoly.*

We call it the Federal Reserve Banking System [The FED] which is a credit/debt system nationally organized by the Federal Reserve act of 1913. This private bank has an exclusive monopoly in money creation in the U.S. All local banks are members of the Fed system, and are regulated by the Federal Deposit Insurance Corporation (FDIC).

6. *Centralization of the means of communication and transportation in the hands of the State.*

We call it the Federal Communications Commission (FCC) and Department of Transportation (DOT) mandated through the ICC act of 1887, the Commissions Act of 1934, The Interstate Commerce Commission established in 1938, The Federal Aviation Administration, Federal Communications Commission. There is also the postal monopoly and Amtrak. [The postal monopoly has been weakened by private companies like FedEx and UPS.]

7. *Extension of factories and instruments of production owned by the State, the bringing into cultivation of waste lands, and the improvement of the soil generally in accordance with a common plan.*

We call it The Desert Entry Act and The Department of Agriculture. As well as the Department of Commerce and Labor, Department of Interior, the Environmental Protection Agency, Bureau of Land Management, Bureau of Reclamation, Bureau of Mines, and the IRS control of business through corporate regulations.

8. *Equal liability of all to labor. Establishment of Industrial armies, especially for agriculture.*

We call it the Social Security Administration and The Department of Labor. The National debt and inflation caused by the communal bank has caused the need for a two "income" family. Woman in the workplace since the 1920's; the 19th amendment of the U.S. Constitution; the Civil Rights Act of 1964; affirmative action.

9. *Combination of agriculture with manufacturing industries; gradual abolition of the distinction between town and country by a more equable distribution of the population over the country.*

We call it the Planning Reorganization act of 1949 , zoning (Title 17 1910-1990) and Super Corporate Farms, as well as Executive orders 11647, 11731 (ten regions) and Public "law" 89-136.

10. *Free education for all children in government schools. Abolition of children's factory labor in its present form. Combination of education with industrial production, etc.*

People are being taxed to support what we call 'public' schools, [which are really 'government' schools because they are owned and controlled by the government] which train the young to work for the communal debt system. We also call it the Department of Education, the NEA and Outcome Based "Education." [4]

The reader is left to decide how threatening the various statutes are to our freedom. Clearly some of them are necessary, and equally clearly the FED is not needed. That is a Jewish bank started by Jews and continues to be owned, run and controlled by Jews. That may be seen as a rather outrageous statement, but consider this.

Mr. R.E. McMaster, the publisher of a financial newsletter called "The Reaper," was able to determine who the Fed's principal owners were through his Swiss and Saudi Arabian contacts. According to McMaster, the top eight stockholders are:

Rothschild Bank of London
Warburg Bank of Hamburg

---

[4] http://www.laissey-fairerepublic.com/TenPlanks.html

Rothschild Bank of Berlin
Lehman Brothers of New York
Lazard Brothers of Paris
Kuhn Loeb Bank of New York
Israel Moses Seif Banks of Italy
Goldman, Sachs of New York
Warburg Bank of Amsterdam
Chase Manhattan Bank of New York

These interests own the Federal Reserve System through approximately three hundred stockholders, all of whom are known to each other and are sometimes related to one another. [5]

It goes without saying that the eight banks listed above are owned by Jews and somewhat surprisingly only half of them are U.S. banks.

Representative Louis T. McFadden, R. Pa., 1915-1935, served as Chairman of the Committee on Banking and Currency for twelve years making him one of the foremost financial authorities in America. About the Federal Reserve banks, he said the following,

The depredations and the iniquities of the Federal Reserve Board and the Federal Reserve Banks acting together have cost this country enough money to pay the national dept several times over. This evil institution has impoverished and ruined the people of the United States, has bankrupted itself, and has practically bankrupted our Government. It has done this through the defects of the law under which it operates, through the maladministration of that law by the Federal Reserve Board, and through the corrupt practices of the moneyed vultures who control it.

Some people think the Federal Reserve Banks are United States Government institutions. They are private credit monopolies which prey upon the people of the United States for the benefit of themselves and their foreign customers; foreign and domestic speculators and swindlers; the rich and predatory money lenders. In that dark crew of financial pirates there are those who would cut a man's throat to get a dollar out of his pocket; there are those who

---

[5] Gary H. Kah, *En-Route to Global Occupation*, Lafayette, LA, Huntington House, Pub., 1996, P. 13.

send money into States to buy votes to control our legislation; and there are those who maintain international propaganda for the purpose of deceiving us and of wheedling us into the granting of new concessions which will permit them to cover up their past misdeeds and set again in motion their gigantic train of crime. . . .[A]nd the Wilson administration, under the tutelage of those sinister Wall Street figures who stood behind Colonel House, established here in our free country the worm-eaten monarchical institution of the "king's bank" [The Bank Of England] to control us from the top downward, and to shackle us from the cradle to the grave. [6]

Thus, with the formation of the FED, in the area of money and banking, the United States went from a republic to a monarchy.

Since the FED was formed in 1913 and Rep. McFadden made that speech in 1932, think of the additional money that has been siphoned off from the U.S. economy since then. For the record the U.S. national debt in 1932 was $19.5 billion which would be $350 billion in 2020 dollars. And, there were 125 million Americans then compared to 331 million in 2020 when the national debt was 26 trillion dollars.

Mayer Amschel Rothschild (1743 -1812), godfather and founder of the Rothschild Banking Cartel of Europe once said, "Give me control of a nation's money and I care not who makes the laws." Clearly in the United States world Jewish banking concerns have control of our nation's money as they do in virtually all other countries.

Henry Ford who was famous for his quotes once said, "It is well enough that the people of the nation do not understand our banking and monetary system, for if they did, I believe there would be a revolution before tomorrow morning."

It is important to understand what having the control of a nation's money supply means. For example, starting in the period after the Second World War the FED caused inflation where prices went up faster than wages. That eventually meant that it took two wage earners to support a family which in turn meant that families would be small with not more than two children and frequently none.

---

[6] Louis T. McFadden (R. Pa.), speech to congress June 10, 1932, Congressional Record, pages 12595-12603.

True Communism

The dehumanization produced by communism is seldom grasped. We hear of the long food lines in Russia under the USSR as well as the gulag prison system where criminals were indiscriminately mixed in with political dissidents. It was a collection of work camps where people were sent to work and die. It is all too easy to equate that with third world countries where people are simply poor. Poverty is only a symptom under communism. The real malice comes from the ideology itself. Professor Denis Flhey gives a good snapshot of what communism means.

MARXISM AND JEWISH NATIONALISM

. . . The Marxian "idea" is that the mass of the material world produces or evolves men who are pure matter, but who in turn modify and change the matter of which future men are composed by changes in the method of production. Therefore all the matter of the world belongs to all the men equally and can belong to no one in particular. As there is no such thing as personality in our sense, owing to the possession of an immaterial, rational soul, there can be no such thing as a right to own permanently any of the means of production. Man may have things which perish in the use of them, like the animals, but mere animal as he is, he has not the right to own land and productive goods in stable possession. He is simply brute matter evolving from the common earth and returning completely thereto.[59] For Marxists, then, private property in the means of production is always exploitation. Human labor, being the labor not of a person, differing specifically from the animals, but of a mere individual, belonging completely to the society, creates value for the society. Private ownership thus means the confiscation of the labor of others to one's own advantage.

There cannot of course, be any question of a native land (*patria*) in the Catholic sense. Material man works and modifies by his labor the particular portion of matter assigned to him by the State-God, but all our language about continuing the tradition of our ancestors is simply meaningless, bourgeois cant. Man is purely material and in due time, given the correct Marxian education, he will be exclusively concerned with matter and its modifications as he should be. According to *The Communist Manifesto* of Marx "the supremacy of the proletariat will efface all national distinctions." Communists, therefore, will take part in a national struggle only as a matter of tactics. Logically, nationality can be for them only a pretence, for nationality supposes the possession of an immaterial soul. If the Communist State grants entire lingual autonomy it emphatically does not recognize *cultural autonomy and liberty.* The national cultures are allowed to remain national in form, but they must be proletarian (that is, materialistic), in content. Little by little the logic of materialism will tend to wipe out national ideas.[60]

Accordingly, any claim to superiority on the part of the Jewish nation is against the principles of Marxism. This is still more strongly the case when the superiority is set forth in accordance with Talmudic tradition. "The Jewish people," writes the Jew, Bernard Lazare, "is the people chosen by God . . . the only one with which the Divinity has entered into an agreement . . . When the serpent tempted Eve, says the Talmud, he corrupted her with his poison. By the reception of the revelation on Mount Sinai, the Jews were delivered from that evil: The other nations remained subject to it . . . other men are thus inferior to the Jews."

59. The inevitable consequence of materialism is brutality and savagery in social relations. "Even the sphere of economics needs some morality, some sense of moral responsibility, which can find no place in a system so thoroughly materialistic as Communism. Terrorism is the only possible substitute, and it is terrorism that reigns in Russia today." (Pope Pius XI *On Atheistic Communism*). This is the truth contained in the satirical work, *Animal Farm,* by George Orwell (Penguin Books). George Orwell, whose real name was Eric Hugh Blair, fought for the Reds in Spain and hated the Catholic Church, yet it is only the doctrine of membership of Christ, taught by the Catholic Church, that can prevent the diffusion of that savagery which compelled him to flee from Spain to escape being "purged" by his "comrades," the Communist Commissars. Cf. Orwell's book, *Homage to Catalonia.*

60. Op. cit., pp. 228-232. Cf. *The Tragedy of James Connolly,*

pp. 11-16.[7]

As seen above Professor Fahey clearly shows the inevitable dehumanization that pure communism brings about. But, he also shows how communism would do away with Jewish superiority espoused in the Talmud, something not likely to happen. Since communism is predominately a Jewish invention it shows the extreme duplicity of the Jewish people in preaching that life will be the same for all under such a society.

Communism as Seen by Russians

Alexander Solzhenitsyn edited a series of eleven essays written by Russians. They are found in the book *From Under The Rubble*. This was after the communists had controlled Russia for fifty-seven years and some of the essayists still lived in Russia putting their freedom and even their lives at risk to write what they did. In the essay by Igor Shafarevich, *Separation or Reconciliation? The Nationalities Question in the USSR*, we find on page 96 an interesting look at what had happened to the Russian identity under communism.

> The belief, it turns out, that Russia has a historic mission, that she too has something of her own, a new word, to offer the world; or, as the authors (contemporary intellectuals who were the mouth pieces of the Russian leadership) put it, "Russian messianism." This is the sin they call on Russians to repent; this, they say, should be Russia's main aim in the future. Their own stated aim is so to change the nation's consciousness that it dare not imagine its life *has* some aim! What other nation has ever been subjected to such sermons?
>
> Several generations of Russians have been brought up on such horrendous versions of Russian history that all they want to do is to try and forget we ever had a past at all. Russia was the "gendarme of Europe" and the "prison of the people," its history consisted of "one defeat after another" and was always characterized by one

---

[7] Rev. Denis Fahey, C.S.Sp., D.D., Ph.D., B.A., *The Kingship of Christ and The Conversion of the Jewish Nation*, Kimmage, Dublin, Holy Ghost Missionary College, Chapter VIII.

and the same phrase: "the accursed past." [8]

This passage is poignant for us because if one replaces Russia with America today he will find exactly what is happening to us. The question "What other nation has ever been subjected to such sermons?" was applicable to the United States. In case anyone has forgotten, under President Barack Obama we were daily lectured on how terrible we were.

However, the destruction of our past remains on-going and pervasive. Civil War monuments in the south are being torn down. George Washington and Abraham Lincoln are covered with opprobrium. Everything that is good with America is said to be bad in spite of the fact that few places on earth have as fine a living standard and as much freedom as do we, though our freedoms continue to erode with each passing day.

How Communism Succeeds

One may well ask what type of individual would embrace communism seeing that at its core it dehumanizes the person. It takes a person who has succumbed to it and then drawn back and repudiated it to give a cogent answer. In this we are fortunate to have a writer of high standing who practically gave his life in the fight to expose communism after he, himself, managed to pull himself out of it at great risk to his life and the lives of his wife and children. This man is Whittaker Chambers who did all that a man could do to expose communism for what it is. In the nine years prior to writing *Witness* Chambers was an editor for *Time* magazine so he had the skills to produce a work of literature as well as a compelling testament against communism.

Whittaker Chambers grew up in what we would call a dysfunctional family. As a young man he was drawn to communism. It is best if we let him tell us how it happened in his own words from Chapter 3.

Sooner or later, one of my good friends is sure to ask me: How did it happen that a man like you become a Communist? Each time I wince, not at the personal question, but at the failure to grasp the fact that a man does not, as a rule, become a Communist because

---

[8] Alexander Solzhenitsyn, *From Under The Rubble*, English version, Boston, Mass., Little, Brown and Company, 1975, p. 96.

he is attracted to Communism, but because he is driven to despair by the crisis of history through which the world is passing.

I force myself to answer: In the West, all intellectuals become Communists, because they are seeking the answer to one of two problems: the problem of war or the problem of economic crises. . . .

This is equally true, even for men of untrained minds or without the habit of reflection; men who find it difficult to explain to themselves or to others the forces that move them to Communism. For while the susceptibility to Communism varies among men, the problem of war and economic crisis do not vary. . . .

When an intellectual joins the Communist Party, he does so primarily because he sees no other way of ending the crisis of history. In effect his act is an act of despair, regardless of whether or not that is how he thinks of it. . . .

Almost without exception such men and women could have made their careers much more profitably and comfortably outside the Communist Party. For the party must always demand more than it gives. . . .

It is in the name of that will to survive that Communism turns to the working class as a source of unspoiled energy which may salvage the crumbling of the West. For the revolution is never stronger than the failure of civilization. **Communism is never stronger than the failure of other faiths**. . . . (Emphasis added.)

It is the crisis that makes men Communists and it is the crisis that keeps men Communists. For the Communist who breaks with Communism must break not only with the power of its vision and its faith. He must break in the full knowledge that he will find himself facing the crisis of history, but this time without even the solution which Communism presents, and crushed by the knowledge that the solution which he sought though Communism is evil against God and man. . . . [9]

Mr. Chambers managed to start at Columbia College in New York but soon dropped out. After fumbling around with his life he returned to

---

[9] Whittaker Chambers, *Witness*, Washington, D.C., Regency Gateway, 1952, pp. 191-193.

study art. In 1923 he and a friend from college went to Europe for the summer to see the art works of the great masters first hand. Again from Chapter 3 he continues.

> I was one of those drawn to Communism by the problem of war. . . . I saw the galleries and museums. But I also saw something else. I saw for the first time the crisis of history and its dimension. It was not only that Germany was in a state of manic desperation, reeling from inflation, readying for revolution while three Allied armies occupied the Rhineland and refugees flooded back from the occupied area into the shattered country. It was not only the aftermath of the World War, the ruins of northern France or what Bernanos would presently call "those vast cemeteries in the moonlight." What moved me was the evidence that World War II was predictably certain and that it was extremely improbable that civilization could survive it. . . . It seemed to me that the world had reached a crisis on a scale and of a depth such as had been known only once or twice before in history. [10]

Upon returning home from Europe, Chambers returned to college and began reading everything he could find on communism.

> I quickly passed on to *Lenin's State and Revolution* and the *ABC of Communism* [the three authors of the latter work were all shot during the Great Purge]. Here was no dodging of the problem of getting and keeping power. Here was the simple statement that terror and dictatorship are justified to defend the socialist revolution if socialism is justified. Terror is an instrument of socialist policy if the crisis was to be overcome. It was months before I could accept even in principle the idea of terror. [11]

He finally made the choice to join the Communist Party not because of the party as such but because he saw no alternative to the crisis he perceived in the world.

---

[10] Ibid., pp. 193-194.
[11] Ibid., p. 195.

> The ultimate choice I made was not for a theory or a party. It was—and I submit that this is true for almost every man or woman who has made it—a choice against death and for life. . . . It demanded of me those things which have always stirred what is best in men—courage, poverty, self-sacrifice, discipline, intelligence, my life, and, at need, my death. [12]

The book, *Witness*, is well worth one's time to read. It is as timely now as when it was written. In its pages it is easy to see that it will take only one major crisis in our time to allow communism to flourish like a wildfire. Notice how he saw in *Lenin's State and Revolution* and the *ABC of Communism* the bold reality that terror and dictatorship are justified as an instrument of socialist policy. How few people would give the slightest thought that terror and dictatorship were in their future if they do not relent in their desire for more and more socialism.

Antonio Gramsci

Few people in the West today think they have a revolutionary socialist/communist future, but they would be wrong on one word. That is, it may not be "revolutionary." Marx and Lenin saw it was a movement of history that the proletariat would rise up in an armed revolt against the capitalists and produce the "Workers' Paradise." When Karl Marx read Darwin's book he immediately thought that communism was the social equivalent of evolution in nature; all societies were naturally evolving to a point where workers ruled all. That has never happened unless one wants to say that republics are a movement in that direction. However, republics are associated with capitalism and free market economies two things communists flatly reject.

Antonio Gramsci (1891-1937) was born in Ales on the island of Sardinia. He went to Turin University in Italy. Gramsci was not Jewish but by 1913 he fell in with the predominately Jewish Italian Socialist Party in Turin. Without going into a detailed development of his thinking suffice it to say he soon saw, as few others did, that some of the basic tenants of communism as espoused by Marx and Lenin were wrong. Here we will

---

[12] Ibid., p. 196.

leave out Stalin because even though he was a communist on some level and he saw it as a movement of history he never believed in a workers paradise. He saw communism as his ticket to become tyrannical ruler of the whole world being a megalomaniac of the first class.

> Marxist though he [Gramsci] was, and as fully convinced as Lenin that there was a force completely inner to mankind driving it on as a whole to the Marxist ideal of the "Workers' Paradise," Gramsci was too aware of the facts of history and of life to accept other basic and gratuitous assumptions made by Marx, and accepted unquestioningly by Lenin.
>
> For one thing, Marx and Lenin insisted that throughout the entire world, human society was divided into just two opposing camps—the broad "structure" of the great mass of people, the workers of the world, and the unjustly created "superstructure" of oppressive capitalism.
>
> Grimace knew otherwise. He understood the nature of Christian culture, which he saw as still vibrant and thriving in the lives of the people all around him. Not only did Christianity point unceasingly to a divine force beyond mankind—a force outside and superior to the material cosmos. Christianity was also the spiritual and intellectual patrimony held in common by the bone-poor peasants in his native Ales, the workers in Milan's factories, the professors who had taught him at Turin University, and the Pope in his Roman splendor. [13]

Once again, it is important to remember that communism is totally materialistic and atheistic; it says there is no spiritual soul in man, there is nothing that transcends the material world we see. Gramsci could see that the simplistic idea that there were two completely separate parts of society—the huddled masses and their oppressors—was wrong. Religion, especially Christianity, united at least on some level, all classes of society. "A key element of Gramsci's blueprint for a global victory of Marxism rested on Hegel's distinction between what is 'inner' or 'imma-

---

[13] Malachi Martin, *The Keys of This Blood*, New Your, Simon and Schuster, 1990, p. 244.

nent' to man and what man held to be outside and above him and his world—a superior force transcending the limitations of individuals and of groups, both large and small." [14]

That is, Gramsci insisted that to achieve the Marxist goal it was necessary to Marxize the inner man. The goal must be a stealthy revolution in the way people think. After all, what is called the Russian Revolution of 1917 was in reality a coup d'état. The people in St. Petersburg were out in the streets rioting, not to overthrow the government, but for fuel and food—they were cold and hungry.

> In the most practical terms, he needed to get individuals and groups in every class and station of life to think about life's problems without reference to the Christian transcendent, without reference to God and the laws of God. He needed to get them to react with antipathy and positive opposition to any introduction of Christian ideals or the Christian transcendent into the treatment and solution of the problems of modern life. . . .
>
> It was in fact pure and simple, therefore, that the residue of Christian transcendentalism in the world had to be replaced with genuine Marxist immanentism.
>
> It was also obvious that such goals, like most of Gramsci's blueprint, had to be pursued by means of a quiet and anonymous revolution. . . . Everything must be done in the name of man's dignity and rights, and in the name of his autonomy and freedom from outside constraint. From the claims and constraints of Christianity, above all. [15]

While Stalin and then Khrushchev followed by Brezhnev, Andropov and Chernenko seemed to be still of the mind that communism would take over the world by the violent uprising of the workers, the rest of the communists of the world were marching to the drum of Gramsci. Mikhail Gorbachev was the first Russian premier to see the writing on the wall. That is, the whole of Western culture was open to the subtle erosion by the immanent, self-centered horizontal view of earth-centered life. Tar-

---

[14] Ibid., p. 247.
[15] Ibid., p.251.

gets of first choice were education from grade school through university, the media, political parties, the family, and religion, particularly Christian religions, and of these especially the Catholic religion.

And, the Catholic Church was ripe for a godless invasion. Already by 1960 there was a well established movement to "modernize" The Church. On the surface it was made to look like the idea was to get rid of some of the trappings the had their origin in the middle ages and before. That is, to bring The Church up to date. For example, among the priest's vestments for saying Mass there was the maniple worn on the left forearm. It had it's origin as a handkerchief but had become stylized and made part of what every priest wore at every Mass. Surely the maniple could be removed. However, that was not what the movement was about. The goal was not so much to "update the externals," but to get The Church "in step" with the modern world, to have it embrace modernism making it essentially a humanitarian charity.

When Pope John XXIII announced the Second Vatican Council he said he wanted to throw open the windows and let in some fresh air. That is not what happened and he knew from the start that was not the goal. Various committees worked for years preparing the documents that would be the guide for the Council. At the beginning of the first session those documents were voted out so as not to be a hindrance to what followed. John XXIII had it within his power to reverse the vote but he did not.

Pope Paul VI showed he was also deeply involved in this nefarious enterprise in his closing address to Vatican II. "So pointedly did the Pope elaborate on that theme of the Church's devotion to subserve material human interests that Gramsci himself could not have written a better papal script for the secularization of Roman Catholic institutions or for the de-Catholicization of the Roman Catholic hierarchy, clergy and faithful." [16] Specifically his closing remarks included the following.

> This **secular religious society**, which is the Church, has endeavored to carry out an act of reflection about herself, to know herself better, to define herself better and, in consequence, to set aright what she feels and what she commands. . . .
>
> Never before perhaps, so much as on this occasion, has the

---

[16] Ibid., p.259.

Church felt the need to know, to draw near to, to understand, to penetrate, serve and evangelize the society in which she lives; and to get to grips with it, **almost to run after it, in its rapid and continuous change**. . . .

The modern world's values were not only respected but honored, its efforts approved, its aspirations purified and blessed. . . .

Any careful observer of the council's prevailing interest **for human and temporal values** cannot deny that it is from the pastoral character that the council has virtually made its program. . .

Would not this council, then, which has **concentrated principally on man**, be destined to propose again to the world of today the ladder leading to freedom and consolation? Emphases added. [17]

Notice that it was never the mission of Christ or the church he founded to "run after society," or to be centered on "human and temporal values." And, to say "The modern world's values were not only respected but honored" is totally at odds with the Christian message which has always been precisely the opposite, to be counter society, to be centered on God and the life beyond this temporal life.

As a parting remark, Paul VI knew precisely what he was saying because before the above parting speech he confided in a close friend that "I am about to blow the seven trumpets of the Apocalypse." [18] And all succeeding popes made no effort to turn back the many abuses and loss of faith caused by Vatican II—Gramsci reigns supreme in The Church as well as the rest of society.

This digression into what happened to the Catholic Church at the Second Vatican Council shows how deeply Gramscian thought had penetrated into society even as early as the nineteen-sixties.

In a later section on Covid-19 it will be obvious that the world has entered the end game where the ultra-rich and hence ultra-powerful cabal has taken control of the world where they will control every aspect of every life.

---

[17] https://www.vatican.va/content/paul-vi/en/speeches/1965/documents/hf_p-vi_spe_19651207_epilogo-concilio.html

[18] It is not possible to go into the source of that remark here. But one can find a detailed account as to why it is credible at https://www.Traditioninaction.org /Questions/F031_TrumpetsApocalipse.html

Chapter 19

Major False Flag Operations

Here will be given some examples of how The Jews, in the meaning given at the beginning of this Part III, that being when acting as a loosely organized body, have affected recent history and continue to do so. The first thing that must be stressed is that the Jews, by design, normally hide their involvement in events that are less then edifying and cause other individuals, groups or nations to appear responsible. This is what are called "false flags." That is, they "plant" the flags of others at the scene of the events to direct attention away from themselves. A surprising number of terrorist acts around the world are known to have been perpetrated by Jews when the true facts are learned. Even at that, it makes little difference how many false flags they plant because of the extent to which they control the media throughout the world. We will start and end this section with significant historical events that changed the lives of millions that at the times they were committed were portrayed by the media as having been done by diverse groups but which are now known for certain to have been perpetrated by the Israeli Mossad (moh'-sahd). In between the major events some smaller events will be presented to demonstrate that False Flag operations of all sizes are the work of the Mossad.

Before going farther it must be emphasized that the Israelis are by no means the only people who have ever tried to deflect attention from the black operations that they conduct. It's just that controlling what the public knows as they do, the temptation to hide their activities is ever present.

The Assassination of John F. Kennedy

A false flag operation of huge proportions is the assassination of President John Fitzgerald Kennedy in 1963. It will be discussed partly because it is an event that most people have heard about and partly because it shows the depth of the Jewish involvement in the world of today. As such it will be explored in some depth to show that, indeed, it was the Jews and in particular the Israeli Mossad that was behind the assassination of Kennedy. The Mossad is Israel's intelligence organization that, like the CIA has two functions. One is collecting intelligence and the other conducting covert operations around the world. Beyond the Mossad one must confront the fact that to plan and carry out a conspiracy of historical proportions it requires a considerable amount of money. To that end it will also be revealed that the largest banking conglomerate in the world the Jewish Bank of Rothschild was the source of much of the funding.

If the Mossad was responsible for the assassination what about the wide variety of other suspects that most of us have heard about for decades. Following is a list and it is hoped that the reader will not be too put off if his favorite perpetrator is missing.

- Lee Harvey Oswald, acting alone
- The Russian KGB
- Fidel Castro
- Anti-Castro Cubans
- The Mafia
- Rogue elements of the CIA acting alone or with others
- The CIA as an official operation
- J. Edgar Hoover and the FBI
- Lyndon B. Johnson
- Texas Oil Tycoons
- The Military-Industrial Complex
- Drug cartels

Before we go any further it is important to establish the facts of JFK's assassination and there are few.

1. Fact: It does appear certain that John F. Kennedy was assassinated in Dallas, Texas on November 22, 1963 by being shot one or more times with a rifle.

2. Fact: No one actually saw anyone shoot Kennedy.

3. Fact: Lee Harvey Oswald was arrested and accused of the assassination by shooting Kennedy from an upper floor of the Dallas Book Repository building.

4. Fact: Jack Ruby shot and killed Oswald on national television so millions of people saw it as it happened.

5. Fact: JFK was killed by a bullet passing through his head from front to rear and Oswald was in the back of Kennedy at the time of the shooting.

Beyond the simple facts listed above there is little else known for certain about the assassination. Many books have been written showing an overabundance of circumstantial evidence and drawing various conclusion therefrom.

In solving any crime it is standard procedure to ask who had means, motive and opportunity to commit the crime. In this case it leaves two basic options. The first is a lone person who knew he would be caught and punished. Or, it had to be an extremely large and sophisticated organization that not only knew it would not be caught and punished but could control the investigating teams as well as public opinion through the media.

An example of the lone operator is the attempted assassination of President Ronald Reagan on March 30, 1981 by John Hinckley Jr. In this case Hinckley knew he would be caught and sent to prison. He was caught and openly admitted his motivation was to impress actress Jodie Foster on whom he had an obsession after seeing the film *Taxi Driver*.

In this light let us first look briefly at Oswald. Many people will argue with Fact 5 above including the Warren Commission. However, years after the assassination the senior surgeon in the emergency room at Parkland Hospital where they brought Kennedy immediately after the shooting, Charles A. Crenshaw, wrote a book about the incident. He delayed his book until he was ready to retire because bad things were happening to people who spoke out with views opposed to the official narrative.

> Then I noticed that the entire right hemisphere of his brain was missing, beginning at his hairline and extending all the way behind his right ear. . . . I also identified a small opening about the diameter of a pencil at the midline of his throat to be an entry wound. I had seen dozens of them in the emergency room. At that point, I new that he had been shot at least twice. [1]

Crenshaw describes how providential it was that Kennedy should be brought to that particular hospital because it was located near a rough part of Dallas. Few emergency rooms in the U.S. had a staff as familiar with seeing gunshot victims as that one. Even though the president's injuries were too sever for even them to be of assistance, they were able to provide expert witness as to the exact nature of his wounds.

> From the damage I saw, there was no doubt in my mind that the bullet had entered his head through the front, and as it surgically passed through his cranium, the missile obliterated part of the temporal and all the parietal and occipital lobes before it lacerated the cerebellum. The wound resembled a deep furrow in a fleshly plowed field. Several years later when I viewed slow-motion films of the bullet striking the president, the physics of the head being thrown back provided final and complete confirmation of a frontal entry by the bullet to the cranium. [2]

In addition, in the brief time Oswald lived after the assassination he claimed he had not done it and that he had been set up as a patsy. No one ever came up with a credible motivation for Oswald's actions other than he must have been under the control of others who even decades later

---

[1] Charles A. Crenshaw, *JFK Conspiracy of Silence*, New York, Penguin Books USA Inc., 1992, p. 78.
[2] Ibid., p. 86.

remained ethereal.

Michael Piper offers a reasonable motive for Oswald, at least as reasonable as any other that has been put forward. Oswald was part of the anti-Castro community and in this narrative he was told he had been selected to be part of a plot to oust Castro.

> There was to be an attempt on the life of President Kennedy so 'realistic' that it's failure would be looked upon as nothing less than a miracle. Footprints would lead right to Castro's doorstep, a trail that the rankest amateur could not lose. Unfortunately for Oswald, he fit the bill perfectly for Hunt's operation." [E. Howard Hunt was a CIA operative.] At first Hunt did not tell Oswald what his exact mission was, except it was of the highest National Security priority . . . It was only two months before the 'fake assassination' when Hunt gave Oswald the rifle, explaining his part in the plan. Oswald was to fire three shots from his rifle 'in the air.' He was to abandon it and empty cartridges at the scene and quickly leave the building for a rendezvous with agents who'd transport him to a secret destination. He would remain in hiding until after Cuba was invaded by the U.S. [3]

Imagine Oswald's horror when he saw that as he shot his rounds over Kennedy's head there were others actually killing him. From the above it seems reasonable that Oswald acting alone did not assassinate Kennedy.

In the case of a super-crime of international scope one must go beyond the simple questions of means, motive and opportunity especially regards motive. One must ask who gains not only in the short term but in the scope of future history.

Israel's Drive to Obtain Atomic Bombs

On the other end of the spectrum from a lone operator is the largest organization in the world, the international Jewish community of which Israel and its Mossad are a significant part. As when discussing anything dealing with the Jews, it is not that there is too little information but quite

---

[3] Michael Collins Piper, *Final Judgement, The Missing Link In the JFK Assassination Conspiracy*, American Free Press, sixth edition, 2005, p. 292.

the opposite, there is so much information that it is hard to decide how to proceed so let us start with atomic bombs.

When JFK took office in 1961 he inherited a plethora of problems, one of which was Israel and its drive to obtain nuclear weapons. It was the early nineteen-sixties and the Cold War between the United States and Russia was the ruling conflict in the world. The club of nations that had nuclear weapons was still small and the threat of proliferation was upper-most in the minds of the leaders of the nations in the club. It was seen that if every small dictator could get "the bomb" the chance of one mistakenly exploding and leading to all out nuclear war between the superpowers was great. And, Israel was a very small country that had no need for such a capability.

The United States had been a strong backer of the State of Israel from its foundation. In fact, the U.S. was the first nation to recognize it as an independent nation when it was formed in 1948. That led the leaders of Israel to assume they could have nuclear weapons if they wanted them. However, JFK saw it differently. Kennedy had already had many troubles with one nuclear power, namely the USSR. Shortly after his inauguration he was handed the vexing problem of the already planned invasion of Cuba where Russia planned to establish a base only miles from Florida. The failed Bay of Pigs invasion was a scar on his reputation as a leader only months after taking office. To add to his dislike for the nuclear equipped Russia, Nikita Khrushchev pummeled him at the Vienna Summit shortly thereafter. Then in October of 1962 there was the nearly disastrous Cuban Missile Crisis. In short, he had had his fill of dealing with nuclear powers.

Israel was building a nuclear plant in the desert near Dimona. They said it was for the purpose of producing electricity but the U.S. was worried they intended to use it to make nuclear weapons. Israel had previously asked the U.S. for help in the weapons endeavor, but both Eisenhower and Kennedy refused to share nuclear secrets, technology and material with them. On May 18, 1963, Kennedy sent a harsh letter to Israeli Prime Minister David Ben-Gurion through the U.S. Ambassador to Israel, Walworth Barbour, saying if they persisted in their efforts to develop nukes the U.S. would abandon their support of Israel. This letter is available but a key part of it is still redacted. The letter from Kennedy to Ben-Gurion will be reproduced here in spite of the missing part because in it he gives the reasoning for not wanting Israel to have nuclear weapons. It shows that he was worried about unrestrained nuclear proliferation. It was not a case of being mean to Israel.

Dear Mr. Prime Minister:

I welcome your letter of May 12 and am giving it careful study. Meanwhile, I have received from Ambassador Barbour a report of his conversation with you on May 14 regarding the arrangements for visiting the Dimona reactor. I should like to add some personal comments on that subject.

I am sure you will agree that there is no more urgent business for the whole world than the control of nuclear weapons. We both recognized this when we talked together two years ago, and I emphasized it again when I met with Mrs. Meir just after Christmas. The dangers in the proliferation of national nuclear weapons systems are so obvious that I am sure I need not repeat them here.

It is because of our preoccupation with this problem that my Government has sought to arrange with you for periodic visits to Dimona. When we spoke together in May 1961 you said that we might make whatever use we wished of the information resulting from the first visit of American scientists to Dimona and that you would agree to further visits by neutrals as well. I had assumed from Mrs. Meir's comment that there would be no problem between us on this.

We are concerned with the disturbing effects on world stability which would accompany the development of a nuclear weapons capability by Israel. I cannot imagine that the Arabs would refrain from turning to the Soviet Union for assistance if Israel were to develop a nuclear weapons capability - with all the consequences this would hold. But the problem is much larger than its impact on the Middle East. Development of a nuclear weapons capability by Israel would almost certainly lead other larger countries, that have so far refrained from such development, to feel that they must follow suit.

As I made clear in my press conference of May 8, we have a deep commitment to the security of Israel. In addition this country supports Israel in a wide variety of other ways which are well known to both of us. [4-1/2 lines of source text not declassified].

I can well appreciate your concern for developments in the UAR. But I see no present or imminent nuclear threat to Israel from there. I am assured that our intelligence on this question is good and that the Egyptians do not presently have any installation comparable to Dimona, nor any facilities potentially capable of nuclear weapons production. But, of course, if you have information that would support a contrary conclusion, I should like to receive it from you through Ambassador Barbour. We have the capacity to check it.

I trust this message will convey the sense of urgency and the perspective in which I view your Government's early assent to the proposal first put to you by Ambassador Barbour on April 2.

Sincerely,

John F. Kennedy [4]

Shortly thereafter Ben-Gurion resigned in no small part because he could make no progress with Kennedy on the nuclear issue. He was replaced by Prime Minister Levi Eshkol so on July 4, 1963, Kennedy sent a letter to Eshkol regarding visits to Dimona saying much the same thing he had to Ben-Gurion and this letter is available in its entirety. What follows is the telegram to Ambassador Barbour at the Embassy in Israel transmitting the text of a letter from President Kennedy to Prime Minister Eshkol.

Dear Mr. Prime Minister:

It gives me great personal pleasure to extend congratulations as you assume your responsibilities as Prime Minister of Israel. You have our friendship and best wishes in your new tasks. It is on one of these that I am writing you at this time.

---

[4] Sources: *Foreign Relations of the United States*, 1961-1963: Near East, 1962-1963, V. XVIII.

You are aware, I am sure, of the exchanges which I had with Prime Minister Ben-Gurion concerning American visits to Israel's nuclear facility at Dimona. Most recently, the Prime Minister wrote to me on May 27. His words reflected a most intense personal consideration of a problem that I know is not easy for your Government, as it is not for mine. We welcomed the former Prime Minister's strong reaffirmation that Dimona will be devoted exclusively to peaceful purposes and the reaffirmation also of Israel's willingness to permit periodic visits to Dimona.

I regret having to add to your burdens so soon after your assumption of office, but I feel the crucial importance of this problem necessitates my taking up with you at this early date certain further considerations, arising out of Mr. Ben-Gurion's May 27 letter, as to the nature and scheduling of such visits.

I am sure you will agree that these visits should be as nearly as possible in accord with international standards, thereby resolving all doubts as to the peaceful intent of the Dimona project. [3-1/2 lines of source text not declassified. However, the redacted lines have, in this case, been recovered and are as follows in bold text:] **As I wrote to Mr. Ben-Gurion, this government's commitment to the support of Israel could be seriously jeopardized if it should be thought that we were unable to obtain reliable information on a subject as vital to peace as the question of Israel's effort in the nuclear field.**

Therefore, I asked our scientists to review the alternative schedules of visits we and you had proposed. If Israel's purposes are to be clear beyond reasonable doubt, I believe that the schedule which would best serve our common purposes would be a visit early this summer, another visit in June 1964, and thereafter at intervals of six months. I am sure that such a schedule should not cause you any more difficulty than that which Mr. Ben-Gurion proposed in his May 27 letter. It would be essential, and I understand that Mr. Ben-Gurion's letter was in accord with this, that our scientists have access to all areas of the Dimona site and to any related part of the complex, such as fuel fabrication facilities or plutonium separation plant, and that sufficient time be allotted for a thorough examination.

Knowing that you fully appreciate the truly vital significance of this matter to the future well-being of Israel, to the United States, and internationally, I am sure our carefully considered request will have your most sympathetic attention.

Sincerely,

John F. Kennedy [5]

The following notes are from Secretary of State Dean Rusk to Ambassador Barbour saying why the need for inspections was so urgent.

In conveying foregoing, you should stress that exhaustive examination by the most competent USG authorities has established scheduling embodied in President's letter as minimum to achieve a purpose we see as vital to Israel and to our mutual interests. Scientific reasons for this are that (a) only a visit before criticality can fully establish features of a reactor--this is reason for requested early summer visit which we hope could be this month or next at latest; (b) it is widely known and accepted by knowledgeable international scientific community that, if intended for ultimate production of weapons grade plutonium, a reactor of this size would be operated to burn a single fuel load approximately every six months, whereas for peaceful purposes optimum burn-up time would be about two years--this is what makes it essential that after mid-1964 visits be scheduled semi-annually.
Rusk [6]

From the point of view of international diplomacy the text highlighted by this author was particularly harsh and was certain to produce an equally harsh response. A photocopy of the telegram above is available on the Internet.

Consider the fact that Israel as a nation was surrounded with Arab

---

[5] Ibid.
[6] Ibid.

countries, none of which were friendly with the Jews. Other than international hype raised by influential Jews there was not much backing for Israel especially since between ten and fifteen thousand Palestinians were killed and three-fourths of a million were forcibly driven from their homes in the process of establishing the Jewish "homeland." That not withstanding, the Israeli leaders believed that their survival as a nation depended on more than the possession of conventional arms.

Their paranoia might have been borne out a few years later by two instances where Israel found itself in a war namely the Six Day War in June, 1967, and the Yam Kippur War in October, 1973. In the Six Day War technically Israel started the fight with a preemptive attack on Egypt who had been massing its troops on the border with Israel. In this case the U.S. stayed more or less neutral because at that time Israel was not considered a client state of the United States. The Yam Kippur War was different. Though U.S. aid was delayed, it was this aid that was pivotal in the State of Israel surviving. President Nixon was not in favor of the aid, but Egypt had substantial aid from Russia and it became a Cold War standoff between the super powers so Nixon had no choice but to come to the assistance of Israel.

In any event, the possession of atomic munitions on the part of Israel did nothing to prevent belligerence on the part of its Arab neighbors. True, Israel only managed to put together their first operational nuclear bomb in December of 1966 so if their neighbors were not deterred from aggressive posturing in June 1967 it was reasonable they did not know about the nuclear capability they faced. But, by 1973 the Arab states might well have suspected that Israel had nuclear weapons at its disposal. Yet once again they were not deterred and once again Israel did not use their nuclear capability. That is to say that Israel having atomic bombs had no deterrent effect on their neighbors and Israel saw that using them was not an option in any case.

In spite of the nuclear weapon's lack of a restraining effect and Israel's not wanting to use them it is odd that Israel would still want them so badly. Even mindful of the old saying that when it comes to security *too much is never enough* is that even a plausible reason for the extreme risks they ran in getting "the bomb?"

As a matter of record Israel's desire for nuclear weapons went beyond wanting an additional amount of security. It became an obsession. This is borne out in what was called The Samson Option. This is the subject of the book *The Samson Option: Israel's Nuclear Arsenal and American Foreign Policy* by Pulitzer Prize-winning New York Times journalist Seymour M. Hersh. It must be noted that Hersh was Jewish. The Samson Option refers to the Samson in the Bible where after the Philistines captured Samson he brought down Dagon's Temple and killed himself along with his enemies. For reference, here are the relevant versus about Samson's final act from the Bible.

23 And the princes of the Philistines assembled together, to offer great sacrifices to Dagon their god, and to make merry, saying: Our god hath delivered our enemy Samson into our hands. 24 And the people also seeing this, praised their god, and said the same: Our god hath delivered our adversary into our hands, him that destroyed our country, and killed very many. 25 And rejoicing in their feasts, when they had now taken their good cheer, they commanded that Samson should be called, and should play before them. And being brought out of prison, he played before them; and they made him stand between two pillars. 26 And he said to the lad that guided his steps: Suffer me to touch the pillars which support the whole house, and let me lean upon them, and rest a little. 27 Now the house was full of men and women, and all the princes of the Philistines were there. Moreover about three thousand persons of both sexes, from the roof and the higher part of the house, were beholding Samson's play. 28 But he called upon the Lord, saying: O Lord God remember me, and restore to me now my former strength, O my God, that I may revenge myself on my enemies, and for the loss of my two eyes I may take one revenge. Revenge myself... This desire of revenge was out of zeal for justice against the enemies of God and his people; and not out of private rancor and malice of heart. 29 And laying hold on both the pillars on which the house rested, and holding the one with his right hand, and the other with his left, 30 He said: Let me die with the Philistines. And when he had strongly shook the pillars, the house fell upon all the princes, and the rest of the multitude, that was there: and he killed many more at his death, than he had killed before in his life. Judges, 16:23-30.

The nuclear option was not only at the core of Israeli Prime Minister David Ben-Gurion's personal worldview, but the very foundation of Israel's national security policy. The Israelis were essentially willing, if necessary, to start an all out nuclear war and blow up the world—including themselves—if they had to do so in order to keep from being defeated by their Arab neighbors.

This policy is further amplified by the fact that the Rothschilds—of the House of Rothschild, the largest banking conglomerate in the world—have been Israel's foremost patrons. In addition, Sam Bloomfield, the Montreal liquor baron, and at the time the head of the World Jewish Congress, was among a clique of money kings who bankrolled Israel's secret nuclear arms program. According to Israeli author Michael Karpin, "in Ben-Gurion's eyes, those who donated to it were 'consecrators' helping to build the Holy of Holies for modern Israel." [7]

It is well to note that Israel's desire for nuclear weapons when put on a religious and even spiritual level as they did could be associated rather directly with the idea of the coming of their Messiah bar David who would be a worldly general and leader to conquer their enemies.

In 1963 this issue of JFK's conflict with Ben-Gurion was a secret from both the Israeli public and the American people and remained so for more than twenty years. All of the evidence shows that it was the issue of nuclear weapons that was the primary cause of Ben-Gurion's resignation and the subsequent assassination of President Kennedy.

Whether or not each single fact presented here is true and unassailable or not does not matter because one gets the overall sense of how interconnected are all of the power players in the United States and the world at large. The normal people who go about their daily lives making society work have no inkling of what is happening behind the scenes. But, it makes sense that the billionaires in industry and organized crime as well as powerful politicians and media moguls would all know one another on some level. And, in many cases they would owe their positions of power to one or several of the other power elite.

---

[7] Michael Karpin, *The Bomb In The Basement: How Israel Went Nuclear and What That Means for the World,* New York, Simon & Schuster, 2006, as reviewed by George Perkovich on https://carnegieendowment.org/2006/02/19/sampson-option-pub-18045.

Here is a list of the main power centers who wanted President John Kennedy removed from office.

- First and foremost was Israel in its drive to get nuclear weapons.
- Israel and its allies in the military-industrial complex because the Vietnam war would continue. Kennedy wanted to end it.
- The CIA's autonomy would continue.
- J. Edgar Hoover's FBI would remain intact.
- The Justice Department would stop prosecuting Mayer Lansky and the Mafia.
- The continuing cover for an ever-expanding joint CIA-Lansky drug smuggling operations out of Southeast Asia.

"We see that "Israel's global network had the power to orchestrate not only the assassination of Kennedy, but also the subsequent cover-up." [8] No other organization or country had the resources and the intimate connections inside the U.S. government and mass media to have any hope of pulling off such an audacious operation without a single individual actually being convicted of a crime. And it must be noted that with Johnson as president it was only a few short years before Israel had atomic bombs.

It was almost as if JFK had a death wish. "Considering John F. Kennedy's secret alliance with the mob during the 1960 campaign, his war against Lansky's underworld syndicate was a double-cross that could not be tolerated." [9] Trying to play both sides in a conflict is a sure recipe for disaster.

And Kennedy's alliance with the mob most surely did him into the White House.

Giancana [Sam Giancana was the mob boss in Chicago] also bragged off microphone about his influence. In *My Story*, a 1977 memoir of Judith Campbell Exner, the Los Angeles woman who was sexually involved with Kennedy in the early 1960s while also meeting with Giancana, the mob leader is quoted as telling her,

---

[8] Ibid., Piper p. 295.
[9] Ibid., p. 295.

"Listen honey, if it wasn't for me your boyfriend wouldn't even be in the White House."

Giancana was not exaggerating. In a 1997 interview for this book, G. Robert Blakey, a former special prosecutor for the Justice Department, said that the FBI wiretaps, many of which have yet to be made public, confirmed that the Chicago syndicate used all its muscle to support Kennedy. "There has been a problem with vote fraud in Chicago really since the turn of the century," said Blakey, who obtained access to the wiretaps in the late 1970s while serving as chief counsel for the House Assassinations Committee. The FBI bugs in Chicago, he told me, demonstrated "beyond doubt, in my judgement, the enough votes were stolen—let me repeat that—stolen in Chicago to give Kennedy a sufficient margin that he carried the state of Illinois." [10]

It wasn't only in Chicago where the underworld helped Kennedy but it is most starkly seen there. It comes out that the top man in the Mafia in the United States was not a Corsican nor an Italian but a Jew, Meyer Lansky. And, the Kennedy family had a feud with him going back decades because "of a Lansky-orchestrated hijacking of one of Kennedy Sr.'s illicit whiskey running deals." [11]

In addition, "Kennedy was preparing for a *rapprochement* with Castro's Cuba" [12] which offended "the anti-Castro Cuban community in Miami, New Orleans and elsewhere." [13] Since the anti-Castro Cubans were connected with the Lansky syndicate, the CIA and the Mossad this alone was cause for wide spread dislike for Kennedy by these organizations that operated in the shadows.

President Eisenhower had thought of dismantling the covert operations arm of the CIA but never acted on it. Kennedy was planning to do just that and at the same time withdraw forces from Vietnam. "This would have been a major blow to the so-called 'military-industrial complex' (of which the Israeli lobby was a major component) that stood to

---

[10] Seymour M. Hersh, *The Dark Side Of Camelot*, New York, Little, Brown and Co., 1997, p. 140.
[11] Ibid.,
[12] Ibid., p. 296.
[13] Ibid.

make immense profits from a continuing U.S. presence in Southeast Asia." [14]

J. Edgar Hover had no reason to like Kennedy because he had learned that "Kennedy planned to merge all of the American intelligence agencies—the FBI included—into a single entity under his brother Robert's direction."[15] Even if Hoover were not removed in this merger his power would be greatly reduced. And, Hoover had his own close ties with the Lansky-linked liquor industries and Israel's Mossad.

One of the most surprising things uncovered by Piper was the extent that illegal heroin running from Southeast Asia dictated the continuation of the Vietnam War. By the time of Kennedy's death "the Lansky syndicate had already set up international heroin running from Southeast Asia through the CIA-linked Corsican Mafia in the Mediterranean. . . . We know today that the Mossad has emerged as a major player as a 'middleman' in much of this drug-smuggling activity." [16]

The above is the briefest summary of the players and motivations in the JFK assassination. Clearly, the Mossad-CIA-Lansky combine had the resources and opportunity to carry off such a massive operation. At this point it is left to mention a few additional players.

Jack Ruby

Jack Ruby was "involved in smuggling arms to Israel." [17] In fact, he was "more Mossad than Mafia, quite in contrast to the old legends swirling about Ruby and his alleged Mafia connections." [18] However, "Lansky's West Coast lieutenant Mickey Cohen . . . also had a long-standing link to Jack Ruby." [19] It may be speculated that when it became known that Oswald "the chosen patsy had not been killed before his arrest, as planned, that Ruby was then told it was his responsibility to finish the job." [20]

---

[14] Ibid.

[15] Ibid.

[16] Ibid.

[17] Ibid., p.300.

[18] Ibid.

[19] Ibid.

[20] Ibid.

Earl Warren

Chief Justice Earl Warren was "apprised by the CIA of possible So-viet Communist involvement in the president's murder [so] was pres-sured into covering up what he mistakenly believed to be the truth about the assassination. . . . Pinning the assassination on 'one lone nut' was Warren's way of protecting America's national security. . . . [O]ne of his fellow commission members was former CIA Director Allen Dulles who had, in fact, been fired by JFK. What's more . . there were immense and multiple Israeli (and Jewish) influences on the Warren Commission staff itself—a factor never considered until the release of *Final Judgment.* Additionally, Warren was also under the influence of his close friend, syndicated columnist Drew Pearson, himself an asset and longtime col-laborator of Israel's propaganda and intelligence arm in this country." [21]

Robert F. Kennedy

Senator Robert F. Kennedy's assassination . . . "was a part of the con-tinuing cover-up of the murder of President Kennedy. In the RFK assas-sination, as we have seen, the Iranian SAVAK—a joint creature of the CIA and the Israeli Mossad—was responsible for coordinating the hit on the senator. Robert Kennedy's death prevented the younger Kennedy from ever bringing his brother's killers to justice." [22]

Lyndon B. Johnson

It has come to light that JFK was planning to drop his Vice President, Lyndon Johnson, from the 1964 ticket. The Jews and the Israeli leaders knew that Johnson had a long history of helping the Jews and being helped by them to a point where he could have been in real trouble if he were ever investigated. There seems to be no record where Johnson pub-licly stated that Israel should be allowed to have atomic bombs, but that is to be expected because a vice-president keeps his mouth shut where his views are contrary to the president's. In any case, with JFK dead and Johnson president, he completely reversed the policy of nuclear weapons development on the part of Israel to the point of letting Israel steal a great deal of highly enriched uranium from the U.S. It was from this material

---

[21] Ibid., p. 362.
[22] Ibid., p. 302.

that they fabricated their first nuclear weapons.

In the early 1960s, the Atomic Energy Commission (AEC) began documenting suspicious lapses in NUMEC's (Nuclear Materials and Equipment Corporation) security, inexplicably lax record-keeping, and the ongoing presence of large numbers of Israelis at the plant. In 1962 the AEC [under JFK] considered suspending "classified weapons work" at NUMEC. In 1965 an AEC audit found that NUMEC could no longer account for 220 pounds of highly enriched uranium. In 1966 the FBI opened an investigation—code-named Project DIVERT—and began monitoring NUMEC's management and Israeli visitors. On Sept. 10, 1968, four Israelis visited NUMEC to "discuss thermoelectric devices with Shapiro," according to correspondence seeking official AEC consent for the visit from NUMEC's security manager. Among the approved visitors was Rafi Eitan. After Eitan's visit, 587 pounds of highly enriched uranium was classified as missing. [23]

"The family of Edith Rosenwald Stern, a prominent New Orleans Jewish leader, were key financiers behind NUMEC. Mrs. Stern was the closest friend of Clay Shaw, the longtime CIA asset charged by New Orleans District Attorney Jim Garrison in the JFK conspiracy." [24] The 1991 film, *JFK*, directed By Oliver Stone was based on the trial of Clay Shaw.

In the absence of any eye witnesses to the shooting of JFK and no subsequent confessions we are left with conjecture. In *Final Judgement*, Piper makes a compelling case for it being a web of interests that include most of the usual suspects listed at the beginning of this inquiry, most of which were Jewish or were owned by the Jews when one looks deeply and objectively at the evidence. The driving motive was Israel's fanatical drive to acquire atomic weapons and Kennedy's equally resolute efforts to stop them.

---

[23] https://original.antiwar.com/smith-grant/2010/04/13/americas-loose-nukes-in-israel/

[24] Ibid., Piper p. 9.

Collateral Damage

In the case of JFK's assassination there were enough players that sooner rather than later the real truth would be known. That it has taken until Michael Piper's book—first edition published in January 1994—was due in no small measure to the fact that people closest to the actual crime and later the investigation kept dying under mysterious circumstances. In the first year after Kennedy's death there were fifteen such deaths and between 1963 and 1976 a total of 112 died of other than natural causes.

The last one to die in the first year after the assassination was a woman by the name of Mary Pinchot Meyer. Here the author will indulge himself and comment on this case because he was peripherally involved in the investigation of it.

Eleven months after JFK was assassinated Mary Meyer was murdered in Washington, DC, on October 12, 1964, along the towpath beside the Chesapeake and Ohio Canal that was by then a park. A man by the name of Peter Janney has written a book, *Mary's Mosaic*, describing the story of Meyer's death and the subsequent trial of a black man, Raymond Crump, Jr., accused of the murder. Crump was acquitted so the case remains open. There are still many unanswered questions about Meyer's death and Crump's trial and many of them surround one Lt. William L. Mitchell.

Janney was continuing his investigation and intended to publish a revised edition of his book with additional findings. Janney's interest in Meyer is that his father and her husband were high up in the CIA and the families of the two men socialized. Since Janney's parents were not much into parenting, Mary Meyer became something of a surrogate mother to Janney as he grew up.

In *Final Judgement* Piper mentions Marilyn Monroe and Judith Campbell as women with which Kennedy had widely-publicized illicit relationships but Kennedy had another long running tryst with Mary Meyer that is less well known. This occurred some years after her divorce from her CIA husband. Meyer was well known in the social circuit around Washington and a regular at the White House. She and Kennedy partied and when not between the sheets together used drugs. As a result there was a lot of pillow talk. To make a very long story short when the Warren Report came out Meyer immediately bought and read a copy.

She told a lot of influential people that she had inconvertible evidence that would blow the report out of the water. Unfortunately for her, the wrong people heard what she said and she was murdered to keep her quiet.

At Raymond Crump's trial, William Mitchell, who purportedly worked at the Pentagon, came forward. He said he had met Meyer on the towpath the day she was murdered as he was jogging in the opposite direction. He identified Crump as a man also on the trail at the same time. Mitchell was one of the prosecution's main witnesses. In an attempt to solve the case Janney dug into Mitchell's life. He reliably reconstruct it until after he finished college.

> During interviews with a number of classmates who knew Mitchell at Cornell, some of whom were also in the various Cornell-administered ROTC programs, a story began to emerge of how Mitchell had come to be stationed at Ft. Eustis in Newport News, Virginia. Mitchell was described by many who knew him (though on one knew him well) as "every private," "serious," "bright", "disciplined," "a whiz at math," and "a physical fitness buff."[25]

After graduating from Cornell he spent a year at Harvard and then things become murky. The records said he was in the Army but Janney could not find a single person who actually knew him in the Army. As such he though Mitchell might have been a CIA operative and his Army record was a fabrication. Eventually he located me who most certainly did remember one Lt. Billy Mitchell.

We both went into the Army at the same time as ROTC lieutenants and landed in the Transportation Corps. The first few weeks of our service was an introductory class to instruct us as to what the Transportation Corps did and we met while attending that class at Ft. Eustis, Va. Janney was delighted when he found me. Mitchell was a man who was hard to forget, though not in ways Mitchell's college classmates described him. Nobody has much privacy in the army, and weather or not he was serious, bright, disciplined, a whiz at math, were hard to determine. He cer-

---

[25] Peter Janney, *Mary's Mosaic*, New York, Skyhouse Publishing, 2013, p. 400.

tainly did not seem to be a physical fitness buff. The thing that made him so memorable was that he had a huge problem accepting the authority of the instructors, unless that was a facade.

Janney makes the case that Mitchell was assigned to the Pentagon to work in military intelligence or to use that as a cover to work for the CIA. On the orders assigning me to my duty station after the class at Ft. Eustis were Mitchell's orders, as well, assigning him to the Pentagon, "USA Elm Def Intel Agcy DOD." [26] That cleared up one mystery for Janney, as in, did Mitchell actually spend time in the Army. But, unfortunately it was not enough to find who actually shot and killed Mary Meyer. Janney makes a good case for it having been the CIA and that makes sense after the case Piper makes in *Final Judgement* for the CIA's involvement in the murder of JFK. The reader should keep in mind that most of the top officers of the CIA were Jews or Zionists with direct contact with Israel.

Masada

One final note on the assassination of JFK. Above the Samson Option was mentioned. Lest the reader think it's a bit over the top to think that the Israelis would commit mass suicide rather than capitulate to their enemies, be advised they have done just that in the past. In the century before Christ the Israelites had built a large fortress on a mountain in the south of Israel called Masada. In 4 BC the fortress was captured by the Romans. In a surprise attack in 66AD the Israelites recaptured it. This was the start of the Jewish rebellion that lasted from 66 to 73 AD. It was during this conflict that Jerusalem was captured by the Romans and the temple destroyed in 70 AD. The Romans laid siege to the Masada citadel for two years until finally on April 15, 73 AD, they were at the point of breaking in. Rather than surrender, the remaining 1,000 defenders led by Eleazar ben Jair committed mass suicide. Only seven women and children who had hidden in a water duct remained to tell the tale. Today the word Masada is occasionally used to signify mass suicide.

---

[26] Headquarters, Department of the Army, Washington 25, D.C., 6 September 1963, Special Orders, No. 219.

Chapter 20

Other False Flag Operations

1967, The USS Liberty

On June 8, 1967, occurred a failed Israeli False Flag Attack Against the United States when:

Israel attacked the American naval vessel USS Liberty in international waters, and tried to sink it in an attempt to drag America into Israel's war [that was the Six Day War that lasted form June 5-11, 1967] by blaming the attack on the UAR, that is, Egypt. After checking the Liberty out for *8 hours* and making *9 over flights* with Israeli jets that came within 200 feet of the vessel the Israelis attacked it with Mirage fighter jets, torpedoes and napalm. The USS Liberty suffered 70% casualties, with 34 killed and 174 wounded. The air attack alone lasted approximately 25 minutes consisting of more than 30 sorties by approximately 12 separate planes using napalm, cannon, and rockets which left 821 holes in the ship. Following the attack by fighter jets, three Israeli motor torpedo boats torpedoed the ship, causing a 40 x 40 foot wide hole in her hull, and machine-gunning firefighters and stretcher-bearers attempting to save their ship and crew. More than 3,000 machine-gun bullet holes were later counted on the Liberty's hull. After the attack was thought to have ended, three life rafts were lowered into the water to rescue the most seriously wounded. The Israeli torpedo boats returned and machine-gunned these life rafts at close range.

The Israelis clearly knew it was an American ship, tried to sink it, and tried to frame the Egyptians for the attack, as shown by the following evidence:

- The Liberty was flying a huge, brand new 5 by 8 foot American flag. The weather conditions were ideal to ensure the flag's easy observance and identification, because it was clear and sunny, with a wind-speed which made for a constant rippling motion in the flag.

- The Liberty had a unique profile and did not look like any other ship, since it had more and bigger antennas—including large, high-tech dishes and giant towers—than any other boat in the world (it was an NSA spy ship).

- The Liberty was marked with uniquely American numbering and colors in front.

- The Israeli pilots shot out the Liberty's communications equipment first, and *specifically* jammed the ship's emergency radio signal ... unique to American naval vessels in the 6th Fleet. The ships from other fleets and other nations used different frequencies, which the Israelis did not jam.

- The Israelis used *unmarked* fighter jets and *unmarked* torpedo boats during the attack.

- Recently-declassified radio transcripts between the Israeli attack forces and ground control show that at least 3 times an Israeli fighter jet pilot identified the craft as *American*, and asked whether ground control was sure he should attack. Ground control repeatedly said, yes, attack the vessel.

- The Israeli torpedo boats methodically destroyed all of the Liberty's life rafts one by one (which is a war crime).

- The only reason the Israelis did not successfully sink the Liberty and kill all of its crewmen was that one sailor duck-taped together an antenna—and took many bullet wounds in the process—which enabled an emergency SOS to get out from the Liberty to American 6th Fleet.

- The Israelis later claimed that they mistook the Liberty for an Egyptian vessel. But the Egyptian ship—the *El Quseir*—was an unarmed 1920s-era horse carrier out of service in Alexandria, four times smaller than the Liberty, which bore virtually no resemblance to the Liberty.

- President Lyndon Johnson believed the attack was intentional and he leaked his opinion to *Newsweek*.

By this act of blatant high seas barbarism Israel hoped to bring the U.S. into their war in a way similar to the sinking of the *Lusitania* that brought the U.S. into World War I.

CIA Chief Richard Helms commented as follows: "Yet the ultimate lesson of the *Liberty* attack had far more effect on policy in Israel than in America. Israel's leaders concluded that nothing they might do would offend the Americans to the point of reprisal. If America's leaders did not have the courage to punish Israel for the blatant murder of American citizens, it seemed clear that their American friends would let them get away with almost anything."
–George Ball, U.S. Undersecretary of State at the time, *The Passionate Attachment.* [1]

Or to put it in other words, if you do not lay down, people won't use you for a rug. The U.S. leadership has persistently let our nation be used as a rug by the Israelis. And why is that? It seems that CIA Chief Helms was feigning ignorance of the fact that any reprisal against the Jews of Israel would have to be sanctioned by the Jews in the United States and that would not have happened then or now. It can be hoped that by now the reader is gaining some appreciation of how thoroughly every aspect of American society, and in indeed the world, is controlled by Jews.

1986 Bombing of La Belle Discotheque in West Berlin

Claims were made that there was "irrefutable" evidence that the Libyans were responsible for the bombing of the La Belle discotheque in West Berlin on April 5, 1986. Libya was one of Israel's favorite enemies. A U.S. serviceman was killed. President Ronald Reagan responded with

---

[1] The above material on the USS Liberty is excerpted from a piece By TS on April 17, 2018, on *Washington's Blog.*

an attack on Libya. However, intelligence insiders believed that Israel's Mossad had concocted the phony "evidence" to "prove" Libyan responsibility. West Berlin police director Manfred Ganschow, who took charge of the investigation, cleared the Libyans, saying, "This is a highly political case. Some of the evidence cited in Washington may not be evidence at all, merely assumptions supplied for political reasons." [2]

2011, Norway Attacks

The July 22, 2011, Norway attacks were reported in the international media as being the work of a single lone wolf terrorist, Anders Behring Breivik, against the government and civilian population in Oslo, and a second against a Workers' Youth League-run summer camp. The attacks claimed a total of 77 lives. The first attack was a car bomb in Oslo, the second where a gunman opened fire with semi-automatic weapons on the island of Utoya. It was the deadliest attack in Norway since World War II. Breivik was said to be closely associated with "Al Qaeda" and other radical Islamic groups.

Yet, there is substantial evidence showing that the attacks came in response to Norway's coalition Labor-Socialist government being seen as hostile to the interests of Israel because Norway announced it would pull its aircraft from the NATO campaign in Libya at the end of July. Norway also announced its support for Palestine's sovereignty resolution in the UN General Assembly. In addition, Norwegian police and intelligence were painfully aware of Mossad's past operations in Norway, including the infamous Lillehammer Affair in 1973. In that case Mossad agents shot and killed a Moroccan waiter, Ahmed Boushiki on July 21. The agents had been sent by Israel to assassinate Ali Hassan Salameh the leader of the Black September Organization, a Palestinian group which allegedly carried out the 1972 Munich Olympics massacre, and they mistook Boushiki for their target.

2012, Benghazi

US ambassador to Libya, Christopher Stevens, and three other Americans were killed on September 11, 2012, when rocket-propelled grenades

---

[2] *The Spotlight*, April 21, 1986," *Final Judgement*, p. 84.

were fired on the US consulate in the eastern Libyan city of Benghazi. We were told that the incident took place while a group of people held a demonstration because of an anti-Islam movie produced in the United States. However, there is good evidence to suggest that the attacks against the U.S. consulate in Benghazi were stimulated, paid for and organized by Israel's Mossad and, hence, had little to do with the Libyan people. President Obama did nothing to assist the embattled consulate personnel at Benghazi in spite of repeated pleas for help.

Why would the Mossad do that? The first reason put forward is that President Obama had an operation going where he was transferring arms and soldiers to Al-Qaeda and Hamas. So President Obama let four Americans die to cover his tracks and get him off the hook with Israel. He needed a sacrificial lamb for Israel so Stevens had to die.

There is another reason that is just as plausible. Israeli Prime Minister Benjamin Netanyahu addressed the United Nations General Assembly in New York on September 9, 2012. As Netanyahu was making his travel plans to the UN he notified President Barack Obama the he would come to Washington after his speech to meet with both of the presidential candidates, Obama and Mitt Romney. It was reported on the national television news that Netanyahu had told President Obama that he intended to come to Washington after his UN address for this meeting and that Obama refused to let him come. We may assume that Netanyahu's purpose was to insure that both candidates understood that which ever of them won Netanyahu would expect his support of Israel. In retaliation for the snub the United States had a dead ambassador. That is how they think; this is how they operate.

Chapter 21

9/11/2001 Attacks on the World Trade Center and the Pentagon

Even though this is out of chronological order it is saved for last because there are few world changing events like it and it is recent enough so nearly everyone will remember it first hand. Once again it is our purpose to show that the Israeli Mossad had the resources, motive and opportunity to carry off this heinous act. Here, in a few pages an attempt will be made to summarize the events of September, 11, 2001, and who was behind it.

Christopher Bollyn in his book, *Solving 9-11: The Deception That Changed The World*, shows that here again a history changing terrorist attack was perpetrated by the Israeli Mossad as a False Flag operation. After reading it one can see why a few Arabs that blew in off the desert could not have planned and successfully executed the most technically involved terrorist plot in the history of the world. Bollyn then details a hundred ways how the Mossad did. "What we need to stand up and say is not only did they attack the *USS Liberty*, they did 9-11. They did it. . . it is 100 percent certain that 9-11 was a Mossad operation. Period." [1]

---

[1] Dr. Alan Sabrosky the Director of Strategic Studies at the U.S. Army War College, March 14, 2010 as related by Christopher Bollyn, *Solving 9-11: The Deception That Changed The World*, Published by Christopher Bollyn, USA, 2012, p. 144.

The Events of 9/11

8:46 am	American Airlines Flight 11, a Boeing 767, crashes into the north tower of the World Trade Center.
9:03 am	United Airlines Flight 175, a Boeing 767, crashes into the south tower of the World Trade Center.
9:37 am	American Flight 77, a Boeing 757, crashes into the Pentagon.
9:59 am	The south tower of the World Trade Center collapses.
10:07 am	United Airlines Flight 93, a Boeing 757, crashes in Somerset County, Pennsylvania.
10:28 am	The north tower of the World Trade Center collapses.
5:20 pm	The 47 story World Trade Center building number 7 collapses.

The World Trade Center Construction and Destruction

The official version is that the fire from jet fuel in the planes weakened the structures and caused the two towers to come down. There was no reason for Building 7 to have collapsed having suffered only minor damage.

After the February 26, 1993, attack on the Twin Towers, Chief Engineer for the WTC, John Skilling, said the towers were built with steel core columns set on a six by eight evenly spaced grid. The structure was designed to have a Boeing 707 crash into it at 600 miles per hour and withstand the impact and the fire from the burning fuel. "Our analysis indicated that the biggest problem would be the fact that all the fuel (from the airplane) would dump into the building. There would be a horrendous fire. A lot of people would be killed," he said. "The building structure would still be there." [2] The 767 is not much larger than the 707 and has only five high-mass components capable of taking out a core column verses seven for the 707.

Since 9/11 there have been highrise fires in Shanghai, Beijing, Grozny, Moscow, and two fires in Dubai. None of those buildings fell

---

[2] John Skilling, "Twin Towers Engineered To Withstand Jet Collision," *The Seattle Times*, February 27, 1993.

down. In 2002, J. J. Beitel and N. R. Iwankiw of Hughes Associates, Inc. did a study of all the tall buildings that were either damaged or fell down from internal fires since 1950. The only buildings with steel core structure to collapse were the World Trade Center buildings. That means the buildings were brought down by demolition charges in its base. Minutes after the collapses Van Romero, an explosives expert and former director of the Energetic Materials Research and Testing Center at New Mexico Tech., told the *Albuquerque Journal* "The collapse of the structures resembled controlled implosions used to demolish old structures and was too methodical to be a chance result of airplanes colliding with the structures. [3]

In addition there were "seismic data that showed unexplained spikes occurring at the beginning of each collapse." [4] If that is not enough there were numerous bystanders, including members of the international press, who were as close as two blocks away that reported seeing explosions near the bases of the buildings at the moment the buildings started to come down. [5] Yet, that was never investigated by our "free press" which is controlled by Jews as shown in a separate chapter below.

Since it appears nearly certain the WTC buildings were brought down by demolition charges, what explosive was used to do it? "A research paper written by an international team of nine scientists led by Dr. Steven E. Jones of Brigham Young University analyzes tiny red-gray chips of highly-explosive thermitic material 'characterized as nano-thermite or super-thermite.' These chips were found in five different samples of dust from the collapsed towers." The twenty-five page paper entitled "Active Thermitic Material Discovered in Dust from the 9-11 World Trade Center Catastrophe" was published in the *Open Chemical Physics Journal.* [6]

Thermite is a mixture of powered aluminum and powered iron oxide (rust). Super thermite has at least one of its major ingredients ground to an nano size with particles in the 100 nano-meter range (1 nm = a billionth of a meter). It will also have small amounts of sulfur, carbon and silicon added. In this form it is the most energetic of conventional explosives. If

---

[3] Ibid., Bollyn, p. 12-13.

[4] Ibid., p. 16.

[5] Ibid., p. 13-14.

[6] Ibid., p. 269.

greater energy density is needed nuclear explosions must be used. This concludes the bare bones layout of the most relevant facts related to the 9-11 disaster. Now as we move on to the means, motive and opportunity of the crime we will start with motive.

Motives for 9/11

We are assuming that Israel was behind the 9/11 terrorist attacks. Why would they do such a thing? There were the following reasons. The first is that Israel had two main wars since its founding in 1948, the six-days war in June, 1967 and the Yam Kippur War in October, 1973. After the second war the ordinary Israeli citizens were livid about having to sacrifice so many of their men fighting Israel's wars. To keep their jobs and, indeed their heads, the leaders set about solving the problem. As a result they began planning a false flag attack on the United states that would be blamed on the Arabs. "In 1978, when Israeli intelligence began planning the false-flag terror operation of 9-11, according to the documented comments of the Mossad chief Isser Harel, [Rafi] Eitan was serving as Menachem Begin's 'advisor on terrorism'. . . . He is one of the architects of 9-11" [7] Remember, we first met Rafi Eitan at NUMEC in Pennsylvania in 1968 stealing 587 pounds of weapons grade uranium. Also, note that 1978 is just five years after the Yam Kippur War and that is when "documented comments" were made and are now in public discourse. It is only reasonable to assume that the first suggestion of such a plan came literally days after the Yam Kippur War. The result of all of this is it left Americans dying fighting Israel's wars in the Mid East since 2001.

Bollyn offers another reason for the 9/11 attacks as Israel wanting to get the U.S. into the "War On Terror" in which we would continually be engaged in wars in the Middle East. This reason was nearly, but not exactly, the same as the first one mentioned. The first reason made the United States Israel's champion so a future attack on them would bring the full wrath of the U.S. down on the attacking nations. The War On Terror was a more general unrest that Israel would use to eventually control all the lands supposedly granted to them in the Old Testament that extended from the Nile in Egypt to the Euphrates in Iraq.

---

[7] Ibid., p. 280-1.

Additional motives are that the 9/11 attacks led to the U.S. invading Afghanistan "where opium production is now at record levels after having been nearly eradicated under the Taliban regime." [8] Yet another reason was the desire for "the construction of an American (Unocal) dual oil and gas pipeline from Turkmenistan through Kabul to Karachi, Pakistan." [9] As can be seen, there were reasons enough for such an outrageous act as happened on 9/11/2001.

Opposed to this, one would have to ask how the Arabs who have been blamed for the attacks might have profited and one is left with little that is positive. It is true that the U.S. had been messing around in the Middle East since the 1950s which led to a lot of hatred toward the U.S. However, the scope of the 9/11 attacks was beyond terrorist groups, and had to have been the work of a stable government. In view of how the 9/11 attacks turned world opinion, especially that of the U.S., against the Arabs it makes no sense that any Arab government would have done it.

Further, the most explosive situation in the Mid East was and still is the question of a Palestinian State. If Israel intended to grab all the territory in the Middle East it would mean the eradication of some of the existing states so they certainly would not want an additional state to be formed, they wanted fewer states. Starting in 1979 President Carter brokered the Camp David Accords that achieved piece between Israel and Egypt. In 1991 there was good progress toward resolving the conflict with Palestine at the Madrid Conference. There followed several productive negotiations right up to the Taba Summit in January 2001 where peace was in sight. These talks were stopped due to upcoming elections in Israel. After 9/11 all progress toward a settlement of the Palestinian issue stopped and has gone nowhere since and has, in fact, deteriorated which is what Israel wanted all along. If that were not enough,

Another Israeli who visibly could not contain his joy at the success of 9/11's "Operation Esther" was the once and future Israeli Prime Minister Benjamin Netanyahu when was asked about his reaction to 9/11, [shortly after it happened] he said: "It's very good!"

---

[8] Ibid., p. 41.
[9] Peter Dale Scott, *The Road to 9/11 Wealth, Empire, and the Future of America,* Berkley, California, University of California Press, 2007, p. 167.

Then, catching himself, he added that while it wasn't exactly good, it was certainly good for Israel. [10]

Notice the Israelis even had a name for 9/11, namely Operation Esther. They celebrate this successful Israeli operation each year on the feast of Purim.

> For more than eleven years, [as of 2012] Israel has been wildly celebrating the success of its 9/11 operation against the United States of America. The latest example: Israeli children recently dressed up as the burning Twin Towers, complete with impaled exploding airplanes, to celebrate the bizarre Jewish holiday known as Purim. [11]

Purim commemorates and celebrates the story of Esther from the Old Testament. The story of Esther occurs while the Jews were in the Babylonian Captivity. Assuerus, the king of Persia, requested his queen, Vasthi, to come to his banquet and she refused his order. At that she was dismissed and the king began searching for the most beautiful maiden in his kingdom to replace her. Esther, concealing that she was a Jew, was chosen. Esther's uncle, Mardochai, refused to bend his knee to the king's chief lieutenant, Aman. This angered Aman who convinced the king to send out a decree to have all Jews in the kingdom killed. However, previously Mardochai had discovered a plot against the king and through Esther warned the king who thwarted the plot. Now, Esther entreated the King not to kill the Jews. At the same time the king discovered that it was Mardochai who had revealed the plot against him and made Mardochai his chief lieutenant. Aman was hanged and the decree against the Jews was revoked. In its place a decree was promulgated allowing the Jews to kill 75,000 of their enemies.

Purim is an inexact comparison to 9/11 in that in the book of Esther the Jews did the killing of their enemies. And, Esther did not insinuate herself into the Kings government as the Jews have done in the United States. She was chosen. It is left unsaid whether she would have been

---

[10] http://whale.to/c/israel_celebrates_successful_911.html
[11] Ibid.

out-of-hand rejected as queen if the king had known she was Jewish. However, having been selected to be queen she succeeded in reversing the decrees where Jews were to be killed and instead their enemies were.

This all goes to the point that says Israel desperately needed to have their enemies destroyed—by others if possible—and Americans have been dying fighting Israel's wars since 9/11. Israel also needed to have America's support of Israel strengthened from the point of view that Israel was and is the most despised nation on earth. Now we move on to means and opportunity.

The Planes Used on 9/11

Several airline pilots have stated that it would be nearly impossible to manually fly a large airliner so accurately as to hit a specific tall building. Yet we are supposed to believe that a couple of Arabs who had never actually flown a real airplane—they only trained on simulators—managed to hit the exact center of skyscrapers two out of two times. This means that the allegedly hijacked airliners were really very large cruise missiles that had been preprogrammed to fly a certain course and hit a specific target. Who would have had the expertise and facilities to pull off such an operation?

There is a company called the Israel Aircraft Industries (IAI) now called Israel Aerospace Industries that is wholly owned by the Israeli Defense Ministry. According to its website it is a world leader in aircraft conversion and modernization programs, unmanned air vehicles, and defense electronics. IAI had various subsidiaries in the U.S. that had been working with Boeing since the 1980s to convert passenger planes to freighters and even tankers. In 2000 it converted a dozen Boeing 767s to freighters all in U.S. plants. This meant they had all the drawings and technical information about the airliners from Boeing. It also meant there had to have been several 767s in their pipeline at various stages of conversion. The reasoning is that they made two of these into guided missiles, painted them in the American Airlines and United Airlines livery and substituted them for the real AA and UAL planes on the morning of 9/11.

There are many serial numbered assemblies on any airliner such as engine cores and landing gear which are nearly indestructible. Some of these parts fell in the streets around the twin towers. With the numbers off these parts it would have been possible to determine exactly which airplanes were used that day. Yet, no investigation was ever made of this obvious forensic evidence. And, why was that? At that time Israeli dual-citizenship Michael Chertoff, a Zionist Jew, was Assistant Attorney General for the Criminal Division of the Department of Justice. As such he oversaw the FBI investigation and the collection of evidence. This airplane evidence as well as most of the other evidence concerning the attack was either destroyed or buried far away from public scrutiny. Suffice it to say, IAI most certainly had the means and opportunity to convert 767s into guided bombs.

There is other circumstantial evidence that supports this. The 767s that hit the twin towers both took off from Boston Logan airport. They flew 47 and 49 minutes respectively before crashing precisely where planned. The other two planes were Boeing 757s which is a smaller niche market airplane. There is no mention of IAI converting any of these. That means that it is possible these were actually hijacked after take off and were being flown by the Arabs. Flight 77 that hit the Pentagon took off from Dulles Airport which is just north of Washington D.C. Yet it flew 77 minutes before nearly missing the Pentagon. UAL Flight 93 took off from Newark and wandered around the friendly skies for 86 minutes before finally, hopelessly lost, crashing in Pennsylvania. As a side issue, there was no wreckage nor bodies of Flight 93 ever recovered, but to go into that is beyond the scope of this inquiry.

Errant Planes

It may come as a shock to most Americans but the idea of terrorists hijacking a large airplane and using it as a weapon of mass destruction was seen as a real danger years before 9/11/2001. As a result plans were in place to send up fighter jets to intercept off-course or non-responding planes and if all else failed to shoot them down. The rationale being that the deaths of all on board would be fewer than those killed in a congested place on the ground where the plane may crash and those on board would die in any case. So if you, as a passenger, ever see an F-15 or F-16 off the wing of your flight, know that they are not there to protect you.

That being the case, why were all four planes allowed to go massively off course and nothing was done? There is the short answer and the long answer. The short answer was that there was a several day terrorist training exercise on-going at precisely the time of the 9/11 attacks. In one scenario of the drill an airliner was to be hijacked and flown into a tall building near the Pentagon. Therefore, the radar operators at the Northeast Air Defense Sector of NORAD (NEADS) and the FAA's air traffic controllers had prior knowledge that there would be false radar targets appearing on their scopes as part of the exercise. So, when they saw the errant blips headed down the east coast from Boston to New York they thought they had been injected into the system and were well within the parameters of what they expected. It took precious minutes for the confusion to be resolved before F-16 interceptors were scrambled to intercept the marauding aircraft only to be too late. There are conflicting reports as to whether Flight 93 was indeed shot down and then only after authorities on the ground knew the passengers on board had taken back control of the aircraft. [12]

The long answer to the question of why the off course planes were not intercepted is that there were not just the four planes in question to deal with. Here the long answer will be drastically shortened. "Colonel Marr of NEADS has said: 'I think at one time on September 11, I was told that across the nation there were some 29 different reports of hijackings.'" [13]

Were all of these false radar blips intended as part of a terrorist training exercise? The answer to that question has never come to light. However, there is a software company called Ptech that was started in 1994 and a key person in the development of that company was Michael Goff a Zionist lawyer. "Ptech software is utilized at the highest levels of almost every government and military and defense organization in the country . . . including the Secret Service, the FBI, the Department of Defense, the House of Representatives, the Treasury Department, the IRS, the U.S. Navy, the U.S. Air Force and last but not least the Federal Aviation Administration."[14] The true long story explains all of this. There are multiple sources for this information some of which are contradictory in certain specific ways, but the overall picture shows nearly

---

[12] Ibid., Scott. In Chapter 13, Scott examines this in some detail.

[13] Ibid., p. 218.

[14] Ibid., Bollyn, p. 114-5.

all critical software used by the U.S. Federal Government is supplied by companies either owned by American Zionists or are owned directly by the Israeli Mossad or Israeli Defense Ministry. One may ask why have we allowed this to happen? It is because those making the purchasing decisions are, themselves, Zionists.

Super Thermite

Once the plotters had devised a means of setting the towers ablaze who had the means and opportunity to bring them down, demolition style, into their own foot print. There are only a handful of outfits in the world that had the expertise to do that. First one must find who made the super thermite in quantity because many tons of it were needed to produce the observed results.

Ehud Barak was prime minister of Israel from July 1999 to March of 2001 when he was replaced by Ariel Sharon. After that he came to America to head up SCP Partners. SCP was a holding company that had Advanced Metallurgical Group, NV in its portfolio. AMG, in turn operated subsidiaries that had the combined capability to produce all of the ingredients of super thermite in sufficient quantity. [15] This provides a plausible source of the super thermite.

In order to place super thermite in the towers in sufficient quantities it required Zionists control the security of the Twin Towers. They worked at this for many years. However, by 2001 Kroll Inc. headed by Michael Cherkasky, a Zionist Jew, had the contract for the WTC security making it possible to arrange for super thermite to be implanted where needed.

Burton Fried was president of LVI Services, Inc., a demolition company that reportedly had done some "asbestos abatement" work in the twin towers. LVI also advertised that it did turn-key demolition projects. [16] Added to this is the fact that in the weeks and days before 9/11 there were mysterious things going on in the World Trade Center.

According to World Trade Center employees Scott Forbes, Ben Fountain, William Rodriguez, and Gary Corbett there were power cuts the weekend before 9/11 - disabling not just electricity but

---

[15] Ibid., p. 279-80.
[16] Ibid., p. 373.

also the WTC cameras and locks. Evacuations also took place. People reported seeing men in overalls carrying huge toolboxes and long wheels of cables, with the explanation that Internet cables were being upgraded. This would give opportunity for conspirators to plant explosives in the World Trade Center under the guise of building maintenance. [17]

Since super (also called nano) thermite can be painted or sprayed on it could have been applied by workmen not even knowing what they were doing. It only becomes dangerous when it is dry.

Warnings by Odigo

If all of the above were not enough, there were reputedly thousands of Israelis who normally worked in or near the World Trade Center. Yet, only one was killed in the attacks. At the time there existed a messaging service called Odigo owned by Israeli interests. "Two weeks after 9-11 Alex Diamandis, Odigo's vice president said, 'The messages said something big was going to happen in a certain amount of time, and it did almost to the minute.'" [18] "According to a report in the Jerusalem Post, Israel's foreign ministry had collected the names of four thousand Israelis believed to have been in the area of World Trade Center and the Pentagon at the time of the attack, yet only one was reported to have died in the Twin Towers. . . . it seems evident that most of them got the warning." [19] If only Israelis were warned, it stands to reason that the Israelis were behind the attacks.

Aftermath of the 9/11 attacks

Only a few words will be said about what happened after the 9/11 attacks. One seldom reported issue is that a Jew by the name of Larry Silverstien had taken over the lease of the twin towers and Building 7 two months prior to 9/11. While obtaining insurance for the properties he included a special clause for terrorist acts. Since there were two separate terrorist acts he claimed double benefits. The insurance company, of

---

[17] https://wikispooks.com/wiki/9-11/WTC_Controlled_demolition#cite_note-15
[18] Ibid., Bollyn p. 8.
[19] Ibid., p. 9.

course, disagreed and the case went to court. It was tried before a Jewish judge Michael B. Mukasey. Not surprisingly Mukasey found in favor of Silverstien who walked away with over 4.5 billion dollars after having paid two months on the lease—how about that for profit . . . and chutzpah!

Alvin K. Hellerstein along with Mukasey, both dedicated Zionists, oversaw virtually all litigation concerning 9/11. Of the ninety-six families who chose to bring wrongful death cases for loved ones who died in the attacks, not one was allowed to go to trial due to the maneuverings and threats by these two judges.

It is of note that there was little in the news about the trials concerning the attempt on the World Trade Center in 1993. It so happened that at that precise time there was another high profile trial on-going on the other side of the country, that being of O. J. Simpson for having allegedly murdered his wife. The juicy trial of Simpson dominated the news and the WTC trial disappeared. What a strange coincidence.

In this short summary of 9/11 only a few of the key players have been named. As one researches a topic like the assassination of JFK or the 9/11 attacks, there appears an overarching presence of Zionist Jews and their surrogates. There emerges a web of interconnecting links among many of the same key players that keep popping up in multiple incidents and in various places including, most of all, our federal government. For example Henry Kissinger was indirectly involved in Kroll Inc. that had the security contract for the WTC in 2001. And as surprising as it may seem, there were "long-standing business and personal ties between George W. Bush and the Bin Laden family." [20]

Before we leave the subject of Israeli False Flag operations it might be well to ask how well radical Zionism is wearing on the world at large and indeed on Israel.

The highly-regarded Israeli historian Avi Shlaim wrote and editorial titled 'Is Zionism today the real enemy of the Jews?' Shlaim's comments were published in February 2005 in the *International Herald Tribune*, (IHT) which is read in 180 nations, but not in the pages of the *New York Times*, IHT's parent newspaper published in the United States. Why would the *New York Times*

---

[20] Ibid., p. 12.

censor the viewpoint of a renowned Israeli historian writing about anti-Semitism? What did Shlaim say that the editors of the *Times* thought was unfit for its American readers? [21]

Bollyn asks the above questions with tongue in cheek because what Shlaim said was rather incendiary and not something the Jewish owned *New York Times* would want its American readers to see.

> Israel's image today is negative not because it is a Jewish state but because it habitually transgresses the norms of acceptable international behavior. Indeed, Israel is increasingly perceived as a rogue state, as an international pariah, and as a threat to world peace. This perception of Israel is a major factor in the recent resurgence of anti-Semitism in Europe and in the rest of the world. In this sense, Zionism today is the real enemy of the Jews. [22]

A delayed effect of 9/11 on New York was seen in the Covid-19 pandemic of which more will be said later. The New York metro area suffered ten percent of the world's Covid deaths in the early months of the pandemic. This was for two reasons. The first was because of the subway system without which New York could not function. Twice a day every day tens of thousands of those who work in the city cram into subway cars guaranteeing that when the virus arrived all would be exposed to it. The other is that at the time of the attacks the owners of the World Trade Center were looking at a cost of a billion dollars to remediate the asbestos in the buildings. When they came down on 9/11 there was a great dust cloud filled with asbestos fibers. This meant that a vast majority of those who lived or worked in New York had a preexisting respiratory condition from breathing the asbestos laden dust that lingered for months in the city. That condition was deadly to those who contracted the virus. Being a Jewish town the Jews decided in characteristic fashion that if they had to suffer everybody would suffer so the whole country was locked down at great cost to our economy and freedom.

---

[21] Ibid., p. 189.
[22] Avi Shlaim, "Is Zionism Today The Real Enemy Of The Jews?" *International Herald Tribune*. Quoted from Bollyn, p. 189.

Chapter 22

Major Media Conglomerates

In his book, *Solving 9-11*, cited above, Christopher Bollyn frequently mentions "the controlled media" yet he never says much about who controls it. Here we will look at the state of modern media that includes news organizations such as TV, radio, news papers, and news magazines, as well as movies, music, videos, social media etc. and see who owns them and if that ownership has an effect on content.

In 1984, 50 independent companies owned the majority of media interests within the United States. As of 2017, there were only six big companies that controlled most of the information and entertainment flow to the public. These top six corporations control from ninety to ninety-six percent of all media. The exact number of corporations and the precise percentages are hard to fix because there is a constant changing of who owns controlling interest in this or that franchise. It also depends on if a given company is mentioned only as to its U.S. market or world wide and if one is speaking of size in relation to gross revenues, profits, or viewers/listeners/readers. The media conglomerates listed below, roughly in order of size, are excerpted from http://tapnewswire.com/2015/10/six-jewish-companies-control-96-of-the-worlds-media/

The Walt Disney Company

The Walt Disney Company ranks as the largest or near the top. Its chairman and CEO, Bob Chapek, is a Jew. The Disney Empire includes several television companies: Walt Disney Television, Touchstone Television, Buena Vista Television, its own cable network with 14 million subscribers, and two video production companies. As for feature films, the Walt Disney Picture Group, headed by Joe Roth (also a Jew), includes Touchstone Pictures, Hollywood Pictures, and Caravan Pictures. Disney also owns Miramax Films, run by the Weinstein brothers. When the Disney Company was run by the Gentile Disney family prior to its takeover by Eisner in 1984, it epitomized wholesome, family entertainment. It still holds the rights to Snow White. However, under Eisner the company has expanded into the production of graphic sex and violence. It owns ABC news and ABC's cable subsidiary, ESPN, which is headed by president and CEO Steven Bornstein, a Jew. Disney also has a controlling share of Lifetime Television and the Arts & Entertainment Network cable companies. It owns seven daily newspapers, Fairchild Publications, Chilton Publications, and the Diversified Publishing Group. Its ownership of Marvel Comics puts it head to head with Time Warner in the battle of super-hero motion pictures. Disney also has a 50% stake in the A&E Networks.

Time Warner, Inc.

Also at or near the top, depending on who's counting is Time Warner, Inc. The chairman of the board and CEO, Gerald Levin, is a Jew. Time Warner's subsidiary HBO is the country's largest pay-TV cable network along with TBS and TNT. CNN is owned by Turner Broadcasting which is a division of Time Warner. Warner Music is by far the world's largest record company, with 50 labels, the biggest of which is Warner Brothers Records, headed by Danny Goldberg. Stuart Hersch is president of Warnervision, Warner Music's video production unit. Goldberg and Hersch are Jews. In addition to cable and music, Time Warner is heavily involved in the production of feature films (Warner Brothers Studio) and publishing. Time Warner's publishing division (editor-in-chief Norman Pearlstine, a Jew) is the largest magazine publisher in the country (*Time, Sports Illustrated, People, Fortune, DC Comics*).

Comcast Corporation

Comcast is also a heavy weight contender. From the gross income point of view it was the biggest several years ago, but others have come up. Comcast received its biggest boost in trying to rise to the top by taking over NBC Universal in 2010. Comcast owns MSNBC which rivals

ABC News of Disney. Comcast also does battle with Disney on the movie and theme park with Universal Studios. Comcast is run by Brian L. Roberts, a Jew, who is chairman, president, and CEO.

After the biggest three it becomes murky as to which are the next largest and what assets any particular corporation controls.

News Corporation

Rupert Murdoch's News Corporation owns Fox Television and 20th Century Fox Films. Murdoch is a Gentile, but Peter Chermin, who heads Murdoch's film studio and also oversees his TV production, is a Jew.

American Amusements

American Amusements is a holding company that presently (2018) controls Viacom, Inc. and CBS.

Viacom, Inc.

Viacom, Inc. is the company that, quite literally, owns the music television game, with control over MTV, VH1 and BET. Viacom also counts Nickelodeon, Spike and Comedy Central among its assets. This big player is headed by Sumner Redstone, a Jew. Viacom produces and distributes TV programs for the three largest networks. It produces feature films through Paramount Pictures, headed by Jewess Sherry Lansing. Its publishing division includes Prentice Hall, and Pocket Books.

CBS Corporation

After spinning off into its own company from Viacom in 2006, CBS has mainly controlled the network television and radio broadcasting side of the business. CBS owns a 50% stake in the CW network, which it shares with Time Warner, while it fully owns ShowTime, CBS Radio and Simon & Schuster, one of the big four of print publishers. CBS also owns its own sports network. Nevertheless, the man appointed by Lawrence Tisch, Eric Ober, remains president of CBS News, and Ober is a Jew.

Twenty-First Century Fox, Inc.

As of June 2018 Twenty-First Century Fox, Inc. runs Fox Entertainment

Group, Inc. as a subsidiary. Peter Chermin, who heads the film studio, Twentieth Century Fox, also oversees the TV production, is a Jew. Its news division, the fair and balanced Fox News ranks as the number one cable news network in America. Also included in the group is the Wall Street Journal, the largest paper in America and the *Times* of London. By sometime in 2019 it is expected that Twenty-First Century Fox, Inc. will be purchased by Disney after spinning off several subsidiaries.

The New York Times

Presently the *New York Times* is its own company with Arthur Ochs Sulzberger, Jr., as the paper's current publisher and CEO. The executive editor is Max Frankel, and the managing editor is Joseph Lelyveld, all Jews. The Sulzberger family also owns, through the *New York Times* Co., other newspapers and magazines as well as three book publishing companies. The New York Times News Service transmits news stories, features, and photographs from the *New York Times* to 506 other newspapers, news agencies, and magazines.

Social Media

Social media is heavily controlled by Jews. Google, Yahoo, Myspace, YouTube, Facebook, and Wikipedia are a few of the large ones.

The above rendering of media giants and their holdings is incomplete as anyone can see, but it includes most of the major players. In addition, the ownership of news and entertainment entities is extremely fluid as already mentioned so the above collections of media outlets will not remain as shown for long. However, it seems that most of the same players, mostly Jews, are always involved. As to whether the collection of the prime media sources in the hands of Jews does in some way mean that they all act in concert is not directly provable. However, it is obvious that there is no longer a major film studio that consistently turns out family entertainment like the original Disney did. In the same way, in 1980 Ted Turner, a Gentile, formed CNN as an independent news outlet. In 1985 he made a bid to take over CBS that sent shock waves throughout the industry. It did not happen. In 1996 Time Warner bought CNN where the public is now safe from the corrupting influence of its former independent thinking and unbiased reporting.

What is also obvious today is the lack of any real news being reported in the major newspapers and television programs. As shown above in the section on False Flag Operations there was an abundance of information available about those events that was systematically ignored, suppressed or outright lied about by means of what is presently known as Fake News.

International Financial Elite

There are a few international financial dynasties that control most of the world's banking and currencies. Anyone who tires to untangle that gets into conspiracy theories that lead to such organizations and agencies as the Illuminati, Freemasons, Bildersbergs, The Trilateral Commission and others until there seems no possibility of getting to the bottom of things. However, if a researcher is observant he begins to see certain names appear again and a again. Among them are JPMorgan Chase, Morgan Stanley, Citigroup, Bank of America, Rockefeller, Warberg, Goldman Sachs, Lehman and others, including overseas institutions such as Deutsche Bank, Lloyds, HSBC, and the City of London bankers who all in one way or another follow the Rothschilds.

> The real villains at the heart of our economic and cultural life are the dynastic families who own the Bank of England, the U.S. Federal Reserve and associated cartels. They also control the World Bank and International Monitory Fund, IMF.
>
> England is in fact a financial oligarchy run by the "Crown" which refers to the "City of London" not the Queen. The City is run by the Bank of England, a "private" corporation. The square-mile-large City is a sovereign state located in the heart of greater London. Considered the "Vatican of the financial world," the City is *not* subject to British law. [1]

Some will say that The Bank of England took over the U.S. economy during the administration of Theodore Roosevelt (1901-1909) when J.P. Morgan took control of a quarter of U.S. industry. Others will say it was with the formation of the Federal Reserve Banking System that was set

---

[1] https://www.henrymakow.com/000447.html

up by the Federal Reserve act of 1913. In any case it was predominately Jews from the world wide banking cabal that were involved.

This topic is merely touched upon here as are many others. Many books have been written on this subject so there is no intent to do justice to it here. It is mentioned to note one more unexpected way our lives are controlled by Jews.

Chapter 23

The Spanish Inquisition

Here we will begin another version of what could be called "A Short History Of The Jews" by discussing selected events and movements that were pivotal in forming today's world. We will connect the Spanish Inquisition, the Protestant Reformation, the Freemasons, the American War of Independence, and the Communist Revolution in Russia.

The connection of the Jews to the Communist Revolution in Russia has been amply demonstrated for our purposes above. We will see that the Spanish Inquisition was primarily aimed at Jews who had agreed to become Christians and were baptized but then either fell back to Judaism completely or maintained their Christianity for outward appearances as they practiced Judaism in secret. Then we will move on the Protestant Reformation and see that it would not have started or at least would not have been as significant as it was if it were not for the Jews and the lack of the Inquisition in Germany. Protestantism, for its part, morphed into Freemasonry in many localities, and the Freemasons were the bridge between the Middle Ages and modern world communism and national socialism.

One of the historical events that is frequently mentioned but little understood is the Spanish Inquisition. When someone wants to hurl an accusation against the Catholic Church they frequently will blurt out, "Well, how about the Spanish Inquisition?" Most Catholics will say nothing because they know as little about it as the accuser. Before we move on, we note that it was not just in Spain that the Inquisition existed and it is more properly called the Medieval Inquisition.

To begin with, the Inquisition was fashioned after the system set up by Moses in the Old Testament. The rules Moses set up are described in Deuteronomy 17: 2-13. He had the problem where his people were constantly in danger of falling away from the teaching of Yahweh and worshipping false gods. Other than the pollution of the Chosen People and their mission, this intrusion of foreign faiths into their society had the unsettling effect of causing disharmony to the point of mob violence and ultimately civil war.

Starting in the second millenium after Christ there was in France the heresy of the Manichees, who were followers of Mani. They were oriental dualists that believed in two gods, one good and one evil with a measure of Gnostic Christian doctrine mixed in. They believed the Great Architect designed the future paradise with little thought of how anyone would ever achieve it. The Manichees believed Adam and Eve were the result of intercourse with the Devil. While Adam was light, Eve was darkness which led to the essential inferiority of women. Marriage, since it propagated the curse of life, was evil. A pregnant woman was seen as possessed by a devil. Suicide, since it ended the crime of living, was good. These and similar tenets led to disruption of society and murders to say nothing of mob violence. And, this was only one of many heresies bedeviling the otherwise settled Christian society of the time.

Efforts were made to counter Manicheeism as well as other heresies but without much success mainly because the heretics were led by wealthy people who had the propensity of buying off or threatening judges. Finally Pope Gregory IX who served as Pope from 1227 to 1241 hit on the idea of putting the Dominican order in charge of the judicial system in which heretics were tired. These men took the vows of obedience, poverty and chastity so were less likely to be open to bribes or threats.

At that early time most of these heretics were not Jews, and even if they had been they would not have been the target of the Church. It had always been the policy of the Catholic Church to leave honest Jews alone to practice their faith however they saw it. However, "the very presence of unbelieving Jews in various parts of Christendom made certain a dissenting minority, intelligent and irreconcilable, and, as certain Jewish modern writers have noted, [provided] a constant nucleus around which dissident elements in the Christian ranks could be assembled." [1]

Gregory IX sent the inquisitors to Spain in 1238. Many of these Dominicans were assassinated as they went about their work but they persisted. The procedure of the Inquisition judicial system was always basically this. Upon coming into a locality the friars would publicly proclaim that all guilty of offenses against the faith must appear and renounce their errors. After that an inquiry would begin. If two witnesses came forward and testified that someone was a heretic that person was put on trial. The whole purpose of the Inquisition from start to finish was to get heretics to recant, take such punishment as deemed justified and be reunited with the Church, which was the prevailing society at the time.

The names of the witnesses were scrupulously kept secret. This was to prevent them from being murdered in retaliation. To compensate for this the accused was allowed to make a list of his enemies. Any witness appearing on the enemies' list was dismissed. Today it is acknowledged by all honest historians that the veracity and learning of the Inquisitorial judges, the rules of evidence, and court procedures were all superior to the civil courts of the day to the point "that men lodged in the civil jails for various crimes would sometimes pretend to be heretics that they might be transferred to the well-lighted, and well-ventilated houses in which the prisoners of the Holy Office were usually kept, and be judged by the Dominicans rather than the civil officers." [2] These simple facts are scrupulously avoided by detractors of the Catholic Church.

The practice where "the names of the witnesses were scrupulously kept secret" has relevance today in the U.S. In high profile cases if the identities of witnesses are not kept secret the witnesses are put into the Witness Protection Program run but the U.S. Marshals' Service at tremendous expense. Lacking our resources in the Middle Ages the reasonable thing was done—testimony was taken in secret.

It must be stressed that "The Inquistor, unlike the judge under Roman Law—the prevailing civil law of the time—was seeking not to punish the offense, primarily, but to heal and reconcile the offender. As a priest, he was not interested in the prosecution of crime—that was the state's

---

[1] William Thomas Walsh, *Characters of the Inquisition*, New York, MacMillan Publishing Company, 1940, p. 22.
[2] Ibid., p. 163.

business." [3]

We must pause here and answer the obvious question many modern readers probably have. Why was it any business of The Church what a person believed? After all, today in the Western world anything goes. You can believe anything you want. Yes, that is true, now as then. It is when one acts on his beliefs that it matters. If today in America a person goes into a crowded area and burns a Quran he will not get far. You may secretly dislike Moslems and that's your business, but do not try to act on it. The same goes for a person's attitude toward other minorities. The point is, there must be a measure of order and safety to society or all that's left as anarchy and that profits no one.

In the year 711 Moors, the Moslems from North Africa, came across the Straits of Gibraltar into Spain. What followed was called the longest crusade that lasted until 1492 when the last of the Moors were finally expelled from Spain. Through all those centuries Spain was always partially under the control of Islam and what remained free of them was a confusion of warring fiefdoms. During that time many Jews came to Spain. A few converted to Catholicism and among these can be counted some of the greatest saints. However, most converts were what were called *Conversos*. These were Jews who had outwardly accepted the faith of Christ and been baptized but remained Jews, in spirit, religious practices and business conduct. It got to a point where there were many priests and even bishops that belonged to this fifth column of *Conversos.*

It was fairly obvious that Jews had sought to build, in Spain, as elsewhere, a new Jerusalem. . . they had tended to work together to this end, offering too, such resistance to assimilation as one might expect in a people once chosen and set apart. Their remarkable success had produced in them a certain hardness, a certain pride. [4]

What hurt them most in the estimation of Christians was the general belief that behind the mysterious cohesion and unity of conduct and purpose observed among them there existed a central control, in the inner circles perhaps of a modern Sanhedrin, which

---

[3] Ibid., p. 105.
[4] Ibid., p. 139.

directed a conspiracy to build up a Jewish State within the state. [5]

That leaves unsaid exactly what the Jews did that made the Catholics dislike them. That is answered as follows.

Unhappily the popular hatred against the Spanish Jews was not wholly without cause. They were disliked not for practicing the things that Moses taught, but for doing the things he had forbidden. They had profited hugely on the sale of fellow beings as slaves, and practiced usury as a matter of course, and fragrantly. As Lea notes, they demanded forty percent interest at Cuenca during the famine of 1326, when farmers needed money to buy wheat for sowing. They were much given to proselytizing, even by a sort of compulsion; thus they would force Christian servants to be circumcised, and urged their debtors, sometimes, to abjure Christ. [6]

In the fifteenth century this developed into an intolerable situation. It became a "land ridden by highwaymen, cutthroats, degenerates, usurers, tax-gatherers, charlatans and scoundrels of a thousand varieties." [7]

This situation could not go on indefinitely without an explosion; and unfortunately there were many explosions of the worst possible sort. The mob, seeing the government of Enrique *el Impotente* unwilling to do anything to curb the Conversos, and virtually handing over to them the conduct of both State and Church, took matters into their own hands. In one city after another, just before Queen Isabel came to the throne, the Conversos were put to the sword and their houses burned. [8]

Into this melee stepped Isabel of Castile. At age seventeen she secretly married Ferdinand of Aragon. It was not until 1475 at the age of twenty-three that she was crowned queen of Castile. From then on the

---

[5] Ibid., p. 143-4.
[6] Ibid., p. 142.
[7] Ibid., p. 135.
[8] Ibid., p. 146.

royal couple set about uniting Spain and restoring peace to the land, and that meant as much as anything reinvigorating the Inquisition. It was through the Inquisition and the systematic military defeat of Moslem cities that finally in 1492 she made Spain once again its own nation. In that year the last of the Moors were told they could stay and die, convert to Catholicism or leave. Most left. The same options were extended to the Jews and most of them left to settle in North Africa, other parts of Europe, and as we have seen some migrated to the Ottoman Empire to eventually become known as the Dönmeh Jews. It may seem harsh to simply say to all Jews that they most pack up and leave. Remember, though, that any Jew who had embraced the Catholic faith and was Baptized was no longer considered a Jew so the Conversos, real or fake, could stay.

At that time the Inquisition did not stop but continued on at various levels of intensity depending on the threat. It was revived with some intensity by Philip II of Spain to combat the Protestants as we shall see.

The Jews, in addition to having an abiding hatred for Christians, had a system of secret societies having little to do with religious Judaism, per se. This system of societies traded and transmitted information around Europe. There were always Jews near the center of power in all countries, if for no other reason than they were ready at hand to lend money. It must be born in mind that at the time Europe was Catholic and it was not permitted for Catholics to charge interest on loans. As a result, the Jews were the only game in town when the Crown found itself wanting to spend more money that the taxes provided, hence loans at usury— from Jews. With information about which emperor was borrowing money, and courtly information in general, Jews knew what was about to happen in political matters and could exploit it financially.

Before we leave the Inquisition it is necessary to mention the number of people who were executed—sometimes by burning at the stake—as a result of its decisions since this number is frequently vastly inflated by those inveighing against the Catholic Church and do not know what they are talking about. The Inquisition was at its height during the time of Queen Isabel of Spain.

Pulgar, secretary to Queen Isabel (himself of Jewish descent), says that in her whole reign [1474-1504] 2,000 persons were put to death by the State, after the Inquisition had handed them over as impenitent, relaxed or pertinacious. This number included those convicted not only of heresy, judaizing, blasphemy and other offenses directly religious, but bigamy, sodomy, and certain other crimes, which in Spain were dealt with by the Inquisition instead of the civil courts. This figure in now generally accepted, even by anti-Catholic historians. [9]

That number of executions was against approximately 100,000 that were brought before the courts of the Inquisition making the executions a low percentage. Opposed to this was the fact that it provided a society that was brought back from barbarism to a land where life was worth living.

As we shall see this may be contrasted with what happened in the rest of Europe where heretics were not held in check in that same time period. That sad omission led to Protestantism. If the Inquisition had been in force in Germany as it was in Spain, Martin Luther would have been taken out of circulation before he could do his damage. That not being the case Protestantism ultimately led to the thirty years' war from 1618 to 1648, a war strictly between Catholics and Protestants. In these years of strife a minimum of three million perished. If all those who died as a result of disease and starvation due to the war are included the number climbs to between eight and twelve million. One hates to see anyone executed, but 2,000 is a small price compared to many millions. As a side note, both Adolf Hitler and Joseph Stalin would also have been either imprisoned or executed if the Inquisition had been operating in their time.

---

[9] Ibid., p. 174. Walsh uses as his source *Cronica de los reyes católicos.*

Chapter 24

The Protestant Revolt

The Protestant Revolt is commonly called the Protestant Reformation but in fact it had little to do with reform and was all about revolt. Yes, there were wide spread abuses in the Catholic Church and some twenty-nine years after the Revolt started the Council of Trent was called and through many sessions ending in 1563 clarified doctrines and put an end to abuses. However, the Lutherans and other Protestant sects did not return to the "reformed" Church giving the lie to the idea that they cared anything about reform.

There were two main reasons for the revolt besides abuses. One was that the Catholic Church had become very rich and avaricious men could not help but covet it. Among these men were, as one would expect, many Jews. The other was in the world of ideas. To appreciate what was going on in the thinking of the times it is well to observe the following brief list of, literally, world changing events.

1348   Black Death, 1348-1350

1456   Gutenberg (1400-1468): printed the first Bible with movable type

1492   Columbus discovered America

1517   Luther (1483-1546): 95 theses posted in the church door at Wittenberg

1536   Calvin (1509-1564): publishes *Institutes of the Christian Religion.*

1543   Copernicus(1473-1543): published *On The Revolution of*

Celestial Spheres presenting the theory of heliocentricity making the sun rather than earth the center of the universe

1546   Council of Trent (1546-1563)

1632   Galileo(1564-1642): published *Dialogue Concerning The Two Chief World Systems* "proving" heliocentricity by observing the moons of Jupiter

1687   Isaac Newton (1642-1727): published *Mathematical Principles of Natural Philosophy* reducing motion and gravity to mathematical laws

The plague called the Black Death was not part of this period but it must be mentioned because its effects were still being felt at the time. In those few years, 1348-1350, between one third and one half of the populations in all Europe from Turkey to Iceland perished.

It ruined the old hearty structure of feudalism. It lowered the potential of life everywhere in numbers, in vigor and in productive power. In some places whole monastic communities were wiped out, in others the Bishop and all his Chapter died. . . . [T]he monastic institution, like all other institutions in Europe, was hit in its vitals and the effect of the blow was felt for generations.

Before the plague the external unity of Christendom was maintained, not only by common doctrine and its consequent civilization, but also by a very large governing set of people who might be called in modern terms "Cosmopolitan." . . .[They] had a speech sufficiently general to all for all to use it habitually amongst themselves.

Now the Black Death had among a hundred other effects the effect of separating local diction and habit between province and province. All this went side by side with a growth of national feeling. A slow division of a United Europe into separate nations would have happened anyhow, but the Black Death accelerated it. [1]

---

[1] Hilaire Belloc *How The Reformation Happened*, Rockford, IL, Tan Books And Publishing, Inc., 1928, pp. 25-26.

In a sense the entire structure of society had to be reestablished and in the 167 years from the plague to the start of Protestantism society still had not fully recovered.

In addition, the world of ideas and technology were changing fast. Gutenberg's feat of developing the movable type printing press did not happen in a vacuum. It was all the more spectacular because for the invention to be worthwhile there was now a need for large quantities of relatively inexpensive paper to say nothing of barrels of ink. It took time for the infrastructure to develop. It also required a high level of organization. Each type character had first to be made in large quantities that were identical. Then the type had to be set for a single page or a set of pages—now called a signature—which when folded and placed in the book formed continuous pages. Imagine the stacks of pages for an entire Bible. Then it was necessary to collate all the pages in the correct order and bind them. This was followed by application of the spine and covers. The first printed books demonstrated the coming of technology. It was one of the first known instances where a complicated product was assembled from "mass produced" parts.

Neither the blow to society caused by the Black Death, nor the new ideas and inventions could have caused the rupture in society that happened without an organizing principle that was at war with Christendom. Bibles were not the only books being printed. The printing press made it possible for tracts and books of all types to be made available quickly and at modest cost thus spreading new ideas at a rate unheard of even a decade before. In addition, at the time of Luther there was something called the battle of the books.

All the powers of international Jewry were allied with, if not actually the motive power of, the vast conspiracy which produced the Protestant revolt. In fact, it was the famous Battle of the Books, a dispute over the Jewish Talmud and Kabbala [vs. the Bible], which set the stage for Luther. Erasmus's friend Reuchlin, the defender of the Jewish books, assembled around him a party of revolutionaries without whose instant support the Augustinian monk's [Luther's] voice might have had no more effect than Huss's or Segarelli's. Once the landslide began, the widespread wealth and power of Talmudic Jewry was organized behind the new movements in

England, Germany and France, and probably gave them their permanence. [2]

The printing press made available large quantities of true Bibles but just as well it provided counterfeit ones that contained many heresies. The Talmudic Jews were in the forefront of this even though most of these tainted Bibles were used and distributed by Luther's followers. "In Ferrara, a great Jewish financial center, they printed heretical bibles for distribution in Italy and elsewhere." [3]

In the mid-twentieth century when this author was growing up, there was a persistent charge hurled at Catholics by Protestants to the effect that the Catholic Church discouraged the reading of the Bible by its members and that it had been so all the way back to the middle ages. This was usually put off by Catholics as the need by Protestants to stress Bible reading because of their reliance on *sola scriptura*. This "prohibition," however, did go back to the middle ages. ". . . the vast traffic in heretical and distorted version of the Bible by the Jews of Ferrara and elsewhere was the true reason why the Catholic Church discouraged her children from reading copies of the sacred work in the vernacular, except in approved versions." [4]

Luther believed that man was saved by faith alone and that good works had nothing to do directly with salvation. He believed man was totally depraved and could do nothing except to plead for mercy from God. Since he broke from the Catholic church he necessarily had to jettison its traditions hence his total dependence on the Bible—*sola scriptura*. Without giving it a second though most people assume the idea of *sola scriptura* came from Martin Luther, but it did not. It originated with the Jews at least as early as the eleventh century.

The Israelite conspirators solved the problem [of destroying the Church] by "cutting through the live part" of the branch and including in their heretical programs ignorance of the Traditions of the Church as a source of revelation, and maintaining that the only

---

[2] Ibid., Walsh. P. 217.
[3] Ibid., Jones, p. 268.
[4] Ibid., Walsh, p. 239.

source of the Truth is the Holy Bible. This war to the death against tradition was renewed by the crypto-Jewish clergy, that is, by the worthy successors of Judas Iscariot, whenever they could, repeatedly from the 11th century until now, with a perseverance worthy of a better cause; achieving their first success in the Protestant Reformation. [5]

Soon another character appeared in the drama called the Reformation. His name as John Calvin. Together, Calvin and Luther, produced what is today called Protestantism.

Luther provided the emotional energy that stated Protestantism. John Calvin (1509-1564) provided the intellectual energy. Calvin, a Frenchmen, developed a whole new version of Christianity, which he published in 1536 in a book called the *Institutes of the Christian Religion*. This book spread rapidly through Europe and seemed to give the people a new faith to replace Catholicism.

Calvin taught the total depravity doctrine: that man is totally evil, "an ape, a wild and savage beast," with a load of sin so heavy and inescapable that man can only submit to it. He taught predestination, that man has no free will, but that everything he does, given his sins, are predetermined by God. He taught the doctrine of the Elect: that God chooses certain people, the Elect, whom He will save, the rest being the Reprobates, whom He will damn. The Elect can be recognized by several characteristics: the have faith; they lead sinless lives; they are materially prosperous, since God rewards His chosen people. [6]

In the above we see that the doctrines of Luther and Calvin were not that different except that Calvin came up with the ideas of predestination, the Elect vs. the Reprobates, and the Elect being materially prosperous. Those facts are always accepted, yet no one bothers to ask where they came from. One can see where some of it came from. Everyone sins and

---

[5] Ibid., Pinay, p. 519.
[6] Anne W. Carroll, *Christ the King Lord Of History*, Rockford IL, Tan Books, 1994, p. 219.

through a normal lifetime the totality of his sins could add up to a burden so large as to lead someone to despair. But, Calvin's ideas of predestination, the Elect and prosperity for them does not fit that scheme of things. It does not fit until we learn that in addition to being a Frenchman John Calvin was a Jew, though a crypto or secret one.

As previously mentioned Jews believe only they are the true sons of Adam and as such are the only real human beings making them the Elect and making the Gentile "an ape, a wild and savage beast," the Reprobates. In the section on Christian Zionists and the Talmud we also saw that the Jews think that all the wealth of the world belongs to them thus making the Elect materially prosperous. If that's the case, how do we account for the fact that the vast majority of Calvinists have been Gentiles. The reader should not misunderstand, Calvin did not intend to create a new Jewish religion. On the contrary, he saw that a vastly corrupted Christianity was a step in the direction of destroying Christianity entirely. He was simply using ideas from the Jewish Talmudic bag of tricks and applying them to a fortunate opportunity. As an aside, it is hard to believe that many Presbyterians think of themselves as members of the Reprobate class, or even the elect for that matter.

We were told by the nuns in grade school that Calvinism was the unofficial religion of the United States. Of course, we were taught nothing of the doctrine of Calvin and how it could relate to the society around us. However, as the years went by it became apparent that the general mindset where everyone strived for more and more wealth gave some credence to that statement. This is not to denigrate the hard working people who had come through the Great Depression, had known deprivation and wanted a little extra in case times turned bad again. But, Calvinism had been a founding principle of America. "The link between Calvin and American Presbyterianism is John Knox (1515-1572), an apostate priest." [7]

Presbyterianism in the United States had three beginnings, all due to migrations from Europe. In 1623 the Dutch settled in New

---

[7] John A. Hardon, S.J. *The Protestant Churches of America*, Westminster, Maryland, The Newman Press, p. 198.

York as members of the Calvinist Reformed Church; in 1629 the Puritan refugees from England landed in Massachusetts at Salem . . . and in 1685 Ulster men, or Scottish-Irish, arrived in New Jersey and Pennsylvania under the leadership of Francis Makenie. [8]

As we noted before there were many reasons behind that rupture of European society that is called the Reformation that affects us to this day. One little known reason was the fact that for centuries the Jews in their study of the Kabbala had been trying to develop magic as a way to make heaven on earth much as they did with Messianic politics. Certainly in politics they had, at times, found themselves in control of vast wealth and power, though it is unlikely that, in itself, it brought any of them heavenly bliss on earth or in the hereafter. In the last half of the fifteenth century and going forward the Kabbalist magic began to be bound up with the rise of the new scientific worldview. Much later C. S. Lewis would make this observation.

> There is something which unites magic and applied science while separating both from the "wisdom" of earlier ages. For the wise men of old the cardinal problem had been how to conform the soul to reality, and the solution had been knowledge, self-discipline, and virtue. For magic and applied science alike the problem is how to subdue reality to the wishes of men: the solution is a technique, and both, in the practice of this technique, are ready to do things hitherto regarded as disgusting and impious. . . . [9]

There is nothing wrong with the basic idea of subduing nature to the wishes of men by the application of science and technology as long as it is kept within the bounds of morality. Who could argue with using technology to make life easier and better for millions of people? However, as we saw in the first part on evolution, this seems inevitably to lead to abandoning reason as the controlling faculty of man, and replacing it with the will—the will to power.

Concerning the will to power, it is well to note that when will be-

---

[8] Ibid., p. 199
[9] C. S. Lewis, *The Abolition of Man*, New York, Macmillan Co., 1943. P. 87-8.

comes the controlling influence in a society civilization suffers. Several authors have discussed this as did H. G. Wells as shown in the following.

> H. G. Wells once made a distinction between communities of obedience and communities of will; he thought that the first produced the stable societies like Egypt and Mesopotamia, the original homes of civilization, and that the second produced the restless nomads of the north. . . . The community of will which we call the Reformation was basically a popular movement. . . . And so Protestantism became destructive, and, from the point of view of those who love what they see, was an unmitigated disaster. [10]

---

[10] Kenneth Clark, *Civilization*, Harper & Row, N. Y., 1969, p. 159.

Chapter 25

Freemasons

So far we have seen that curbing the negative effects of the Jews on society was the main driving reason for the Inquisition, that they had a significant role in the Protestant Revolt and that John Calvin was a significant intellectual motive force behind it. What is the step from Protestantism to the Freemasons?

Freemasonry was secularized Calvinism. The Elect became Lodge brothers through the alembic of time. Despite changes in names, the revolutionary distillate was the same in its essence, hatred for the Catholic social order or Europe once embodied in the Habsburgs of Spain but now embodied in the Bourbons of France. [1]

We have mentioned that Spain expelled the Jews in 1492. Well, the English did the same thing in 1290 and the French followed suit in 1306. Many of these displaced Jews fled to Scotland whence eventually came the Scottish Masons.

The standard history of Freemasonry has it starting with the founding of the Grand Lodge of London in 1717. However,

. . . one document showed that the first American Lodge was founded in Newport, Rhode Island in 1658 when 15 Jewish families migrated from Holland. That and the colonization of the Mas-

---

[1] Ibid., Jones p. 508.

sachusetts bay Colony by English Judaizers would ensure America became the "new Jewish world order" in the 20th century. [2]

Jews who studied the Talmud and were into Kabalistic mysticism formed secret relationships among themselves. One of their goals was to find a way to rebuild Solomon's Temple. Since the guilds of masons were in existance from the early Middle Ages, it is reasonable that there was a meeting of the two that resulted in a secret society. That means it is impossible to determine a specific date when the Masons of today really started.

The origins of the lodge are surround in mystery. Masons on the continent invariably traced their history back to the destruction of the Knights Templar in 1326. Whether the Templars fled to Scotland . . . is speculation based on documentation lost in the mists of time. What is fairly certain, however, is that the lodges in Scotland were "operative" before they were "speculative. . . ." Gradually, their penchant for construction took a mystical or metaphysical bent, and Freemasonry became involved in an attempt to recover something great which had been lost. Freemasonry would achieve this through recovery of the name of Jehovah and the reconstruction of Solomon's Temple. [3]

Finding reliable, consistent information on the Freemasons, or what is commonly called simply Masons, is difficult. Part of the difficulty is in separating two meanings of the term freemason. The first is what is called *operative* masonry which means real masons which included architects and engineers as well as masons along with their apprentices and helpers. Any time a king or noble wanted to make a building or fortification out of stone he needed masons. Some historians trace this trade back to the Tower of Babble and before. In the middle ages various trades formed guilds that were similar to what we, today, might call labor unions. These organizations did primarily two things. First they made it possible for these tradesmen to earn a fair, consistent wage. On the other

---

[2] Ibid., p. 479.
[3] Ibid., p. 484.

side they had a rigorous program of apprenticeships so that someone who hired a member of the trades had a measure of assurance that the person was competent. Freemasons were those who had passed the requirements of apprenticeship and became basically free agent itinerant masons traveling about to where there were building projects that would hire them.

The other form of what we call masons are usually called *speculative* masons and have little to do with the trades, though, at the beginnings, as noted, there was some interrelationship. Following the Protestant Revolt cathedral building in Europe all but stopped. There is some dispute about whether or not Jews started the speculative masons, but generally it is accepted that it always was predominately a Jewish organization. From its earliest times it contained many Masonic symbols and rituals that lean heavily to Jewish practices. Many lodges through its history were specifically open only to Jews. The Masons of today are basically a Jewish organization that is worldwide and, though it appears open to the world, it is in reality a secret organization. Here secret is not meant specifically secret in its rituals of the lower orders something that has been widely known. But, to the members of the upper orders who have knowledge of its true inner workings, schemes and designs there is the utmost secrecy.

That Freemasonry is a Jewish organization is undeniable. Following are a few quotes out of many that could be given to demonstrate this.

The following could be read in the French Masonic magazine Le Symbolisme (July 1928): "The most important duty of freemasonry must be to glorify the Jews, which has preserved the unchanged divine standard of wisdom."

The high-ranking freemason Dr Rudolph Klein stated: "Our rite is Jewish from beginning to end, the public should conclude from this that we have actual connections with Jewry." (Latomia, No. 7-8, 1928).

A speech at the B'nai B'rith convention in Paris, published shortly afterwards in The Catholic Gazette (London) in February 1936 and in Le reveil du peuple (Paris) a little later, stated: "We have founded many secret associations, which all work for our purpose, under our orders and our direction. We have made it an honor, a great honor, for the Gentiles to join us in our organizations, which are, thanks to our gold, flourishing now more than ever. Yet it remains our secret that those Gentiles who betray their own and most precious interests, by joining us in our plot, should never know that those associations are of our creation, and that they serve our purpose..."

"Freemasonry is a Jewish establishment, whose history, grades, official appointments, passwords, and explanations are Jewish from beginning to end."----Rabbi Isaac Wise

Ray Novosel, writing from Australia in 2004, states: "Zionist world leaders, men in influential positions with the various Masonic organizations everywhere, have worked "hand in glove" for a universal world revolution, which will bring in the One World Church and a One World Government. Many Masonic Lodges are exclusively Jewish, as are the controlling B'nai B'rith Lodges — the mother of the infamous and very dangerous Anti Defamation League (ADL)." [4]

It is well to define as well as possible what Freemasons believe. Various branches of Freemasonry differ somewhat in there beliefs across the world and through time. Basically it is a naturalistic religion that reverts to deism and pantheism. The general tenet is that all religions are equally good and one is as good as another in ones quest of reaching the Grand Lodge in the sky. However, they have an abiding hatred for what was called altar and throne. By altar was meant the Catholic Church and throne meant the Catholic monarchs of Europe. Both stood for what was called *absolutism* which meant that there was an absolute moral code by which people must live. This they hated. Their goal was through moral license, especially sexual license, to make people think they can have heaven on earth which is the goal of all revolutionaries. In fact Mystical Masonry is synonymous with revolution. In his encyclical Pope Leo XIII said the ultimate aim of Freemasonry is:

10. . . the utter overthrow of that whole religious and political order of the world which the Christian teaching has produced, and the substitution of a new state of things in accordance with their ideas, of which the foundations and laws shall be drawn from mere

---

[4] http://www.whale.to/b/jews_and_freemasonry.html

naturalism. . . .

12. Now, the fundamental doctrine of the naturalists . . . is that human nature and human reason ought in all things to be mistress and guide. [5]

If one accepts that line of reasoning, as the Masons hope people will, then passions will rule and licentious behavior will pervade society leading to lawlessness and anarchy. At that point they will be there to fill the void with tyrannical brutality. Leo XIII continues:

20. . . .For, since generally no one is accustomed to obey crafty and clever men so submissively as those whose soul is weakened and broken down by the domination of the passions, there have been in the sect of the Freemasons some who have plainly determined and proposed that, artfully and of set purpose, the multitude should be satiated with a boundless license of vice, as, when this had been done, it would easily come under their power and authority for any acts of daring.

21. What refers to domestic life in the teaching of the naturalists is almost all contained in the following declarations: that marriage belongs to the genus of commercial contracts, which can rightly be revoked by the will of those who made them, and that the civil rulers of the State have power over the matrimonial bond; that in the education of youth nothing is to be taught in the matter of religion as of certain and fixed opinion; and each one must be left at liberty to follow, when he comes of age, whatever he may prefer. . . .

27. Now, from the disturbing errors which We have described the greatest dangers to States are to be feared. For, the fear of God and reverence for divine laws being taken away, the authority of rulers despised, sedition permitted and approved, and the popular passions urged on to lawlessness, with no restraint save that of punishment, a change and overthrow of all things will necessarily follow. [6]

---

[5] Pope Leo XIII, *Humanum Genus* (The Race of Man—the title is taken from the first words of the encyclical) April 20, 1884.
[6] Ibid.

There is one more aspect of the Masons that should be mentioned. They have a system of life insurance that is lucrative in that it has low premiums compared to the value of the policy. In the past, the one caveat was that if the holder of the policy were a Catholic or former Catholic it was a requirement that he not be allowed to see a priest to receive final confession and last rites when he was on his death bed. If he did, the policy became null and void. Whether or not that is still a practice of the Masons, it was in the past. If nothing else, that should demonstrate the anti-Catholic nature of the Masons.

Chapter 26

Masons in the American War of Independence

There is clear evidence that the Jews helped win the American War Of Independence. Once again, they could not have done it alone, but it might not have happened without their involvement. Though small in number many Jews chose to cast their fate with the Colonies. This was because the land was so vast and the people were so hard at work surviving that the Jews went mostly unnoticed which was fine with them.

As with every other country the Jews settled in, they were not patriots, as such. They came primarily due to persecutions in other countries and to set up shop and take over the economy. However, men such as Aaron Lopez were bankrupted supporting the war when their ships were lost to the British. In the area of finance the young American government might have floundered too except for the support taken on by Hayim Solomon. Solomon was to die bankrupted by his total support of the American cause.

The thirteen colonies had a lot of differences since many had been founded by religious sects persecuted in Europe such as the Puritans. As such, they had very specific ideas about what form of government they should have and these ideas differed substantially.

Eventually Freemasonry crept in and from its ranks came the idea of the American Union. The Grand Master for North America Masonry was Joseph Warren and the Green Dragon coffee house on Union Street in Boston is generally considered to be the site where one of its offshoots

The Sons of Liberty plotted the Boston Tea Party. Some sources say the Boston Tea Party was really a recessed meeting of that Masonic Lodge.

Of the leaders that eventually met to form the United States the only thing they had in common was Freemasonry. Both Samuel Adams and Patrick Henry were Masons.

> Leading up to the American Declaration of Independence there were two fiery orators, Samuel Adams from Massachusetts and Patrick Henry from Virginia. Henry delivered passionate speeches for independence, declaring on one famous occasion (in a phrase which was popular with the Freemasons and liberals in France who would spark the French Revolution): "Give me liberty or give me death." [1]

The American Declaration of Independence was adopted on July, 4, 1776, but another document preceded it, namely *Common Sense* by Thomas Paine that was first published in January 1776. *Common Sense* "became one of the seminal documents of the American republic." [2] That was because in 35 pages it gave a strong argument for the separation of the colonies from England and thereby to form a separate nation.

There are differences of opinion about whether or not Paine was a Freemason, though later in life he was to write a history of the Masons. It does not much matter because he was thoroughly of the age and in *Common Sense* he states that the rationalist view should be the only one by which people should be governed. That is, all laws must come from the people, that governments obtain their powers from the governed and not form God. We have been brought up with the idea that "majority rule" as the only reasonable thing to do. Few people would argue with that unless they have been on the losing end of the majority. That is, the tyranny of the majority over the minority is still tyranny.

Pauly Fongemie researched and compiled information on the intellectual climate prevailing in America and Europe in the latter eighteenth century in which we see again that Freemasons were an integral if not the

---

[1] Anne W. Carroll, *Christ and The Americas*, Rockford, IL, Tan Books, 1997, p. 103.

[2] Ibid., Jones p. 530.

principle part of the thinking that was to ferment the American War of Independence that resulted in the United States. Fongemie states:

> . . . [Benjamin] Franklin was admitted to the Nine Muses [the intellectual center of French Freemasonry in Paris] and became Master of the Lodge. There he devoted himself to a propaganda campaign which swung French public opinion in favor of the American Masonic cause. Franklin's "admirable work," said Fay [French historian Bernard Fay], "was the most carefully planned and most efficiently organized propaganda ever accomplished, and "made possible the military intervention of France on the side of the Americans."
>
> Moreover, he asserted, Franklin's work also had "a great intellectual influence throughout Europe, spreading the idea, or what might be called the myth, of virtuous revolution." Up until that time, the French historian said, revolutions had been viewed "as crimes against society." Subsequently, revolutions "were accepted as a step in the progress of the world," a step and a perception which "originated with the American Revolution and grew out of Franklin's propaganda." [3]

As late as 1781 the war had not been won by the Americans nor was it lost by the British. Arms were being funneled into the Colonies by arms merchants running the British blockade primarily from the tiny free trading Island of Dutch St. Eustatius. (St. Eustatius Island is located about 500 miles east of the Dominican Republic in the Caribbean.) Jewish merchants and arms traders were a major presence on the island.

In 1781, the British saw they had to cut off arms shipments to the rebels through St. Eustatius. Admiral Sir George Rodney was sent to capture the island. His goal was to destroy the supplies and destroy the island's commercial and merchant class so they could not provide any more aid to the Americans. Early in 1781 the lightly defended island fell to the main British battle fleet. Rodney destroyed the warehouses and most of their contents and burnt every home. Jewish property was

---

[3] Pauly Fongemie, *Freemasonry: Foundation of the American Revolution,"* www.catholictradition.org.

confiscated and the men imprisoned. Rodney spent months directing half his fleet to convey the most valuable goods back to England.

While Admiral Rodney was engaged in St. Eustatius, Lord Cornwallis and his army of British regulars were forced out of the Carolinas and retreated to the small port of Yorktown, Virginia, on the James Peninsula where he awaited critical re-provisioning and reinforcements. The weakened British fleet, with Cornwallis's reinforcements, was intercepted at sea by the French fleet under Admiral DeGrasse and defeated. After that Degrasse took up position at the mouth of the Chesapeake Bay blockading Yorktown from the sea. General George Washington saw his chance and besieged the trapped Cornwallis. In short course Cornwallis surrendered. The war was over. The Americans had won with the help of the French and indirectly by the Jews on St. Eustatius.

Freemasonry in the United States continued to grow. On May 31, 1801, the first Supreme Council of the Thirty-third Degree, the Mother Council of the World, declared its existence at its meeting in Charleston, North Carolina. It announced a new 33-degree system of high degrees that incorporated all 25 of the Order of the Royal Secret, and added eight more, including that of 33°, Sovereign Grand Inspector General. This new organization declared control of high-degree Masonry in America. As a matter of interest the number 33 was chosen because the 33rd parallel is the closest parallel to Charleston. The Masons have continued to this day.

There is an interesting observation that can be made. To anyone who thinks about it they will see that nearly anyone they know that is successful in business is a Mason whether or not he is a Jew. Since there are only fifteen million Jews in the world they need helpers. Or they may have thought that if every single successful person was a Jew someone might get suspicious. It would be a corollary to this premise that not every gentile who became a Mason would be successful, but one's chances of success without being one would be next to nil, hence gentiles are admitted to most lodges.

From the above and many other sources one could quote, it can be seen that the Freemasons and thereby Jews were a determining factor in forming what was to become the United States of America. The creed of Freemasonry is the rationalist, naturalist mindset that as a practical matter takes God out of economics and society in general.

Masons in Modern Israel

The rise of Masons to political power in Israel dates back to 1948. David Ben-Gurion, Israel's first Prime Minister, was a Mason. Every Prime Minister has been a high level Mason, including Golda Meier who was a member of the women's organization, the Co-Masons. Most Israeli judges and religious figures are Masons. The B'nai B'rith Lodge of New York is affiliated with Israel's lodges and so is the ADL and ACLU, not to mention almost every top investment tycoon on Wall Street.

Much of the trouble in the world today originates in modern day Israel. As we saw in False Flag Operations above, they are responsible for the most heinous acts of terrorism around the world including the United States, their premier supporter. They are also a major problem to their neighbors in the Middle East. This is in no small part due to the fact that:

> Most of the Jews born and living in Israel are not originally from the Middle East (Mazrahim) or the Mediterranean areas around Spain, Italy, and North Africa (Sephardim). Instead, most of their ancestors come from Central and Eastern Europe. . . . In this sense, they are physically *in* the Middle East but culturally *divorced* to a large extent from the practices of their neighbors who inhabited the land of Palestine for centuries before them.

> What gives Israel such a powerful influence in the Middle East is her close ties to the USA, owing to the fact that several major Jewish families hold generations-long dynasties in United States finance, banking, and industry. By serving as the bankers to governments around the world and using money in the form of debts to be paid with interest (usury), families such as the houses of Rothschild, Rockefeller, Warberg, Goldman, Sachs, Kuhn, Loeb, Lehman, and Seaife were able to use their position as debt holders to bend the will of governments to their personal interests. In the words of Meyer Amschel von Rothschild, founder of the Rothschild banking family, "give me control over a nation's currency and I care not who makes the laws. [4]

---

[4] Andrew Bieszad, Understanding the Modern Middle East, *Catholic Family News*, June 2018. P. 16.

Chapter 27

The Nazi Holocaust

The charge that the German Nazis exterminated six million Jews in what is call the Nazi Holocaust is another of those world events where if not a false flag, has misleading information associated with it that is used to misdirect world opinion to believe something other than what really happen. In what follows we will try to untangle yet another myth perpetrated by "the Jews." Here again there is a massive amount of information about a subject that is available in books but especially on the Internet. And, it is one more case where the average person is so totally convinced of the truth of the allegation that it would not occur to him to question it.

In the Nazi concentration camps we are told that from ten to thirteen million people died. We are also told that half of them were Jews. That would make a nice round number of six million Jews that perished in what is called the Jewish Holocaust. The others who died at the hands of the German Nazis were predominantly Catholics (ten percent of the population of Dachau were Catholic priests) but also many Gypsies. Yet, all you ever hear about are the Jews. In fact, every city of any size in the U.S. has a Jewish holocaust museum.

On the Internet there are several sites that refute the six million number. To begin with in 1939 there were about 15.6 million Jews in the world. In 1947 there were 15.7 million. These are numbers from Jewish sources such as the Statistical Bureau of the Jewish Synagogue Council of America. From these numbers it is not reasonably possible for six million or roughly 40% of all Jews in the world were killed between those two years. In addition, the greatest number of Jews living in the lands Germany controlled at it's greatest expansion never exceeded 4.5 million. As one would expect, when the roundup of Jews started early in WW II Jews began moving to places like Russia in droves. As the cult of six million began to gain footing, as shown in what follows, the world population numbers of Jews changed. Still, eventually there were enough reliable sources to show the six million was not a real number.

To solve this puzzle we start with the fact that the concept of six million Jewish victims of prejudice has been an organizing principle for Jews throughout the world since long before World War II. Or more to the point, it has become one of the ideas that defines modern Jewry.

So, where did the number of six million come from? That is nearly impossible to define exactly but some say it is biblical. The argument goes like this. It will be recalled that there are a group of rabbis that study the Kabbalah in the hope of learning deep truths about God and his creation. One of the esoteric practices they engage in is Gematria. This is the Kabalistic practice of interpreting texts by associating words and numbers to Hebrew characters since there are no numerals in written Hebrew. In so doing they came up with six as a mystical number. There are six days in creation, six points to the Star of David. The Six Day War in 1967 lasted from June 5 to June 11 which is seven days, but six is the mystical number. (Jews will say that the last major military action where they captured the Goal Heights ended on the tenth. But, the cease fire brokered by the UN that ended the war occurred on the eleventh so it was a seven day war.)

Here one source will be cited that seems to summarize much of the information that is available on the subject of the six million. John "Birdman" Bryant has a web page titled *History & Scriptural Origin of the Six Million Number*. In it he states the following:

Regarding the "six million" number you should know the following: In the Hebrew text of the Torah prophesies, one can read 'you shall return'. In the text the letter "V" or "VAU" is absent, as Hebrew does not have any numbers; the letter V stands for the number 6. Ben Weintraub, a religious scientist, learned from rabbis that the meaning of the missing letter means the number is "6 million." The prophesy then reads: "You will return, but with 6 million less." See Ben Weintraub: *The Holocaust Dogma of Judaism*, Cosmo Publishing, Washington 1995, page 3. The missing 6 million must be so before the Jews can return to the Promised Land. Thus, it would seem that the six million is numerologically predetermined and can never change. It must be ritually repeated and publicly acknowledged. [1]

The phrase "you shall return" comes from the book of Joshua 1:15. It is as follows: "Until the Lord give rest to your brethren, as he hath given you, and they also possess the land which the Lord your God will give them: and so **you shall return** into the land of your possession, and you shall dwell in it, which Moses the servant of the Lord gave you beyond the Jordan, toward the rising of the sun." Emphasis added.

From this one can see that the Jewish leaders adjusted their thinking to demand that in some way six million Jews had to be sacrificed before a state of Israel could be formed or more properly reformed. This idea of six million was not new with the WW II, however. That same web page then mentions two times that the idea of six million was used long before the war. They are as follows.

A. On page 482 of the article on 'anti-Semitism' in the 10th Edition of the Encyclopedia Britannica (1902) is found the words: "While there are in Russia and Rumania six millions of Jews who are being systematically degraded...". This reference precedes references to the Six Million of WW2 by approximately 40 years.

B. In the American Hebrew Magazine of October 31, 1919, there appeared an article entitled "The Crucifixion of Jews Must Stop!" By Martin H Glynn, former governor of the state of New York. This article begins, "From across the sea, six million men and women call to us for help . . . ."

To continue with John Bryant's web page we see that there is more to the six million than simply a mystical number. There is a continuing flow

---

[1] http://Thy-weapon-of-war.blogspot.com/2009/02/origins
-of-the-six-million-number.html

of money to the Jews to be had by keeping the Nazi Holocaust of six million alive.

But this is still only half the story, because the Holocaust is not merely a religious dogma intended to weld the Jewish community together in much the same way that the old-time Judaic religion once did, but it is also a fabulous cash cow which has placed billions upon billions of guilt-money into the hands of both the Jewish Establishment and individual Jews. ("There's no business like Shoah [Holocaust] business" is a common saying among the Jewish set.) In particular, even 60 years after the end of WW2, the German government is still paying 'reparations' to Jews -- this for a calamity that never happened to a state that did not exist at the time -- and the United States, whose entry into WW2 sealed Germany's defeat, and whose highest legislative body has often been referred to as 'Israeli-occupied territory' -- has been equally generous to the Jewish state, if not more-so. In short, therefore, the Holocaust lie has been both the spiritual and financial wellspring of modern Jewish power. [2]

Another reference to six million is found in a work by Max Raisin. He says the following:

. . . the day of complete deliverance finally heaves in sight, and amidst the anguish and suffering of a great world war [WW I], the harrowing effect of which no part of the inhabited globe is permitted to escape, mankind experiences a new birth, and the Jew, too, at last is about to come into his own. . . . A new day is at hand when the weak and oppressed of the earth will find themselves permanently delivered. . . . The beginning of this complete emancipation has already been made in the wonderful change in the status of six million Jews wrought over night, as it were, by the Russian Revolution. [3]

---

[2] Ibid.

[3] Max Raisin B.A. LL.D., *A History of The Jews In Modern Times*, New York, Hebrew Publishing Company, 1919, pp. xi-xii.

Here again there were not six million Jews available to be counted at that time in Russia. Even several years later after the USSR had been formed there were only between 2.5 and 4.0 million Jews in all of the countries swept into the USSR and most of them were in the Ukraine that was not part of Russia at the time the above comment was made.

There are several more such references from books and articles in the early twentieth century were we see that whenever a Jewish writer wanted to refer to a large number of Jews, the number six million immediately came to his mind. Also notice that in the first two citations it concerns bad things happening to Jews while in the third it is supposedly a good thing so "six million" is not identified only with calamities.

Another take on the six million Jews comes once more from Max Raisin's book quoted above.

On March 30 [1917] all restrictions [in Russia] were removed from Jews in the army and navy, while on April 5 was issued the general emancipation decree abolishing all limitations of race and religion, and granting to all inhabitants of the land freedom of residence and movement, of property ownership, of commercial, industrial and professional pursuits, of education, of the use of any language or dialect of choice, and of participation in the government. The great Russian revolution has not yet brought to Russia the national happiness which ordinarily forms part of political liberty. . . . Yet whatever the result of the present turmoil and chaos, there is no likelihood that the freedom already gained will ever be lost. . . that Russian Jewry will never again be put in bondage. . . . Political emancipation may not bring much of an abatement of Jew-hatred in Russia any more than it did in Germany or France, and we may even look forward to a renewal of anti-Jewish hostilities with all of their old-time excesses and horrors. To the Russian people, as to the world at large, the Jews will still remain an exotic plant in their midst, one, perhaps. . .to be suspected and feared. . . . In the meanwhile the freedom now possessed by these six million Jews will make for a deepened and intensified Jewishness. . . . Ten months of freedom have only served to make the Jew in Russia

more loyal to his spiritual heritage, more anxious to uphold traditions revered and treasured by his forefathers for thousands of
years, and more determined to make real and permanent the new
life which is opening up to him in the light and spirit of Jewish
nationalism. [4]

The above excerpt if typical of a Jew writing about Jews. Once again
he refers to the mystical "six million" Jews. After that one must ask why
should the Russian people have to fear Jews who have been given political and economic freedom? Why would that even come up? It is because
when the Jews are given full freedom in another society they without fail
form a "Jewish nation" within the host nation? They never join up and
become members of the host nation. Notice, there is no mention of the
Jews becoming good Russia citizens.

This author grew up in a small farming town. Mother's parents came
from Germany, and father's grandparents did. It would happen at times
when mother's relatives visited they would speak German. When they
left mother would rail against them for being so "Deutschy." More than
anything my parents wanted to fit in and be Americans. The Jews are not
like that.

Added to this is the fact that Jews always over play their hand. True,
there is precedent that in more modern times it paid off as it did in them
taking control of Russia in 1917 and in the United States since then.
However through history it frequently got them into trouble or expelled
from the host country.

They began to enter Russia's schools in large numbers; by 1881,
one out of every twelve university students came from a Jewish
family. Yet this apparent betterment of the Jews' situation did not
reflect a change in Russian popular attitudes, and the masses continued to despise them as economic exploiters and Christ killers. [5]

That characterization says it all. Jews, where they are persecuted,

---

[4] Ibid., pp. 209-11.
[5] W. Bruce Lincoln, *In War's Dark Shadow, The Russians Before the Great
War*, New York, The Dial Press, 1983, p. 212.

bring it on themselves because either they are economic exploiters or Christ killers, usually both. An example of this is when they were expelled from Moscow in 1892.

Jews who had lived in Moscow fifteen, twenty-five, even forty years were forcibly removed to the Jewish Pale of Settlement [an area in the Western part of Imperial Russia where from 1791-1917 most Jews in Russia were forced to live]. In all, Moscow lost over twenty thousand Jews, some one hundred million rubles in trade and business production, and about twenty-five thousand jobs that Russians had held in Jewish enterprises. The production of silk, one of the city's most lucrative enterprises, was virtually wiped out. [6]

Notice that Jews owned the silk business "one of the city's most lucrative enterprises." It is typical that if there is a lucrative business they will control it to the exclusion of others. Also, for every Jew—man, woman, and child—they had in their employ on average of more than one Russian. Add to this, the Jews are rarely known as generous employers Then when the native population finally has had enough and drives out the Jews all we hear is the "poor Jews" but never the reason why that happened.

It is important to note that this author is in no way suggesting that the Nazi Holocaust did not happen. From several sources it seems that a good average number is about one million Jews that died in the Nazi camps. That was certainly a tragedy and one can hope it will never happen again. However, it must be kept in mind that many more millions of Ukrainians died at the hands of Jews in the Ukrainian Holocaust than Jews died at the hands of Nazis in the Nazi Holocaust.

There is an interesting thing about population statistics concerning the Jews. Above we mentioned that there were 15.6 million Jews in the world in 1939 and 15.7 in 1947. However, once the Holocaust industry got into high gear, the number of Jews fell to about 11 million by the late forties of course depending on whose numbers you used. As a result

---

[6] Ibid., p.218.

when researching the Jewish population around the time of WW II it is important to use numbers from that time, not present numbers as for example from the Internet.

Another thing the Jews will say when their supposedly six million is compared to the others who died in Germany or in the Ukraine is that the six million is a much larger percentage of all Jews than is true for other races. But, there were between fifteen and sixteen million Jews in the world in 1947 when the world population was 2.5 billion. Today there are still only fifteen to sixteen million Jews when the world population is 7.0 billion. Why don't the Jews reproduce? Or is it they falsify the numbers so everybody will continue to think they are a persecuted race that is barely managing to stay in existance?

In the previous section on Why Communism Succeeds in one of the quotes from *Witness* by Whittaker Chambers he refers to "the Great Purge." This occurred in the years 1935 to 1938 in Russia. It is more rightly called the Great Communist Party Purge brought about by Joseph Stalin because that was when Stalin was solidifying his power. There was, as the numbers suggest, a large contingent of the Communist Party whose idea of what communism should be that differed from Stalin's.

The political purge was primarily an effort by Stalin to eliminate challenges from past and potential opposition groups, including the left and right wings led by Leon Trotsky and Nikolai Bukharin, respectively. Following the civil war and the partial restoration of the Soviet economy in the late 1920s, veteran Bolsheviks no longer thought it was necessary to perpetuate the temporary wartime dictatorship, which had passed from Lenin to Stalin upon Lenin's death in 1924. Stalin's opponents accused him of being undemocratic and lax on bureaucratic corruption. As a result he enforced a ban on those party members who opposed him, effectively ending democratic centralism. In the new form of Party organization, the Politburo, and Stalin in particular, were the sole dispensers of ideology. This required the elimination of all Marxists with different views.

As with all such mass executions and killings there are a wide range of numbers given. Here the number 1.2 million is a commonly used median number. This is not to be confused with the much larger number— 15 to 20 million—who died of starvation or were worked to death in the gulag prison system in the nineteen twenties and thirties.

It is well to point out that most of the Communist Party members in Russia at the time of the Great Purge were Jews so most of those put to death were Jews. Therefore, it is not an unreasonable statement to say that the Jews killed approximately as many Jews in the Great Purge as the Nazis killed in Germany. That is, about one million in each case. And, both purges occurred in the same part of the world in the same ten year time span. Yet, all we ever hear about are the Jews killed in Germany.

As an aside, above it was mentioned that the United States' congress is "Israeli-occupied territory." In this line one should recall that Prime Mister Benjamin Netanyahu addressed a joint session of congress in March 3, 2015. During the speech he was interrupted by 39 applauses and 23 standing ovations. Why would our elected representatives do something like that? And more to the point, why was he allowed to give such an address at all? What other world leaders address our congress? The answer is that the vast majority of men and women who get elected to the House of Representatives or the Senate in some way get money either from Israel directly or from the powerful Jewish lobby. After President Donald Trump was elected in November 2016 and inaugurated in January 2017 there was a continual effort to show that his election was in some way made possible by the collusion of the Russians. If anyone wants to find where the real election tampering in the U.S. is coming from they need look no further than Israel.

The Preservation of the Holocaust Mind Set

It has now been more than seventy years since the end of the Nazi Jewish Holocaust. How is it that each succeeding generation has such a vivid horror of it? The only way, of course, is that it is forcefully instilled into the minds of the young. In this vein, there is a telling piece that can be found on the Internet. It is a trip report by Alice Ollstein who at the time was a senior at Santa Monica, California, High School.

I jumped at my grandfather's invitation to attend the annual American Israel Public Affairs Committee (AIPAC) policy conference in Washington D.C., where I could join more than 1,000 students and 4,000 adults in discussing the future of Israel. I returned from the conference, however, feeling manipulated, disturbed and disgusted with a great deal of what I witnessed there.

The first thing I noticed about the conference, besides the sheer volume of participants, was the carefully manufactured atmosphere of fear and urgency. The cavernous hall that hosted all our meals and plenary sessions was always filled with dramatic classical music, red lighting and gigantic signs reading "Now Is The Time." That, combined with the montages of terrorism footage projected onto six giant screens, whipped the audience into a "Save Israel" fervor that most found inspiring. . . .

The mealtime plenary sessions taught me the most about the mindset of our country and about the art of rhetoric. Each speaker played upon the audience's deepest fears and greatest hopes. Even after four days, it never ceased to amaze me how easy it was to get a standing ovation out of the crowd. Even platitudes such as "I believe in democracy" or, even better, "I believe in Israel" had the crowds on their feet. . . . as Israeli candidate for prime minister Benjamin Netanyahu put it, "The world is split between those who oppose terror and those who appease it." [What about the Mossad that actively engages in terrorism?]

The speakers and "informational" videos left no gray area, no place for dialogue or debate, and certainly no place for dissent. I especially squirmed at the parallels AIPAC drew between Iranian President Mahmoud Ahmadinejad and Hitler. To the tune of more dramatic classical music, the six enormous screens flashed back and forth between Hitler giving anti-Jew speeches and Ahmadinejad giving anti-Israel speeches. The famous post-Holocaust mantra "Never Again" popped up several times. Everything was geared toward persuading the audience that another Holocaust is evident ... **unless we get them first**. . . . I noted the scariest division of all: You're either pro-war or pro-Holocaust. Emphasis added . [7]

The above conference hosted 1,000 impressionable young people, but just as disturbing were the 4,000 adults. Is it feared that without traumatizing young and old alike in an "atmosphere of fear and urgency" that

---

[7] http://www.rense.com. Posted by Alice Ollstein on June 28, 2006

normal Jewish people might decide to join into the rest of society and get along? Do not forget the "billions upon billions of guilt-money [put] into the hands of both the Jewish Establishment and individual Jews" as part of the Holocaust industry.

The above conference is not an isolated event today, and the roots of such behavior go all the way back to the time of Christ.

By rejecting Logos, which is simultaneously the person of Christ and the order of the universe, including the moral order, which sprang from the divine mind, the "Jew" found himself drawn inexorably to revolution. The parent of the man born blind knew "the Jews. . . had already agreed to expel from the synagogue anyone who should acknowledge Jesus as the Christ. . . ." Judaism was a 'religio licita' or permitted religion—a religion freed from offering sacrifice to the emperor. Explosion from the synagogue must have inflicted severe hardship on the Jewish followers of Jesus, because they thus also lost protection and social membership. [8]

The process of selecting who may and may not be members of the synagogue led to making the synagogue a revolutionary cell. The process is easy to see. The fanatics keep becoming more and more extreme and the others could either go along or be expelled from the Jewish community. At the time of Zevi Sabbatai (1666 AD) those who did not go along with the enthusiasm of the time when the Jewish rabbis proclaimed him their Messiah were expelled from the synagogue then, too. In both of these cases the most radical elements managed to pull the entire community in their direction.

The radicals who determined the direction of the Jewish community after Christ's death were known as Zealots. Jews who followed Jesus were expelled from the synagogue, just as now. Jews living in the Diaspora today who do not support the aims of the Likud Party (the Israeli political party favoring terrorism and revolution) in Israel are being rooted out of the Jewish community.

---

[8] Ibid., Jones, *The Jewish Revolutionary Spirit*, pp. 41-42.

Chapter 28

Unlikely Connections

Israel and China

On a related note, there is a newly formed relationship between Israel and China. In 2000 the CIA first learned of a strong relationship between the Mossad and China's equivalent of the CIA, the CSIS. That meant that almost everything that either Israel or China learned about the U.S. was shared with the other.

> It was the call that Tenet [head of the CIA] had been waiting for during the past two months. On that frosty morning in January 2001, Senior Colonel Xu Junping of China's People's Liberation Army, and one of his nation's new breed of military high fliers, a strategist with fluent English and a year at Harvard University on his CV, had defected to the United States. [1]

Early in September of the same year a Chinese delegation went to Kabul, Afghanistan to sign an agreement for China to supply the Taliban with modern weapons in exchange:

> . . . the Taliban would order Muslim Fundamentalists to stop their long-running terror attack against China's western provinces. . . . On Saturday, September 8, [2001], the Mossad had sent a short

---

[1] Gordon Thomas, *Seeds Of Fire*, Tempe, Arizona, Dandelion Books, 2001. P. 478.

transmission "burst" that the previously troubled Chinese western provinces were now quiet—and that a delegation from Beijing was due to Kabul on Tuesday, September 11. [2]

It just so happened that on September 11, 2001, Lt. General Mahood Ahmed, head of Pakistan's intelligence service, PIS, was in Washington soliciting various intelligence services for additional funding.

By lunchtime on that terrible Tuesday, Ahmed's schedule had been cleared for him to join Tenet at Langley. For the rest of the day, Ahmed briefed the CIA director and his senior staff on the link between bin-Laden and China. It matched everything Junping had earlier said during his debriefing. Ahmed told him that China had made a decisive decision: It was prepared to infuriate America and its allies in supporting bin-Laden and the Taliban because Afghanistan fitted into China's own long term strategic plans.

Hours later, Tenet received a coded "red alert" message from the Mossad's Tel Aviv headquarters. The CIA chief was presented with what he called a "worse case scenario" that Tenet had always feared—that China would use a ruthless surrogate to attack the United States. [3]

As we have seen in False Flag Operations above, there is over-whelming evidence that Israel's Mossad was the "ruthless surrogate" referred to in the "red alert" message. This book, *Seeds Of Fire*, was published shortly after the 9/11 attacks so Thomas may have been grasping at a few straws. However, if what he relates is true that message was probably one of the more extreme examples of *chutzpah* ever.

Moslem Terrorism

While not a false flag operation as such, the scourge of radical Islamic jihadists around the world today is largely the work of the Mossad and their Zionist confederates in the CIA. How did this happen? Moslems in the Mideast did not one day suddenly decide to become jihadists. In the

---

[2] Ibid., p.492.
[3] Ibid., pp. 492-3.

graduating classes of the University of Egypt in the nineteen-fifties there were numerous women and hardly a burka to be seem. They were peaceably becoming westernized. That is to say, under Gamal Abdel Nasser Egypt had become for all intents and purposes a secular state. This went on until communists backed by the USSR began to encroach on the Middle East, especially Egypt, and here again we see the Jews at work. Just as the Russian revolution could not have happened if the Jews had not had the disgruntled, or rather, starving, Russian peasants to tap into, they needed other surrogates as their foot soldiers in the Mideast and jihadists were perfect.

It is no exaggeration to say that the problem with jihadists today is the work of the CIA and the Mossad. "The then leader of the Egyptian Muslim Brotherhood, Sayed Kuttub, a man Faisal [Crown Prince of Saudi Arabia] sponsored to undermine Nasser, openly admitted that during this period [1960s] 'America made Islam.'" [4]

The Central Intelligence Agency has always been heavily controlled by Jews and Gentile Zionists. The CIA was being formed at about the time as the State of Israel came into existance. President Truman signed the law forming the CIA on July 26, 1947 and Israel officially became a State in May 14, 1948. One cannot say the two are essentially linked but it shows the wide activity of the Jews after the war especially in America.

The Afgan-Russian War from December 1979 to February 1989 was the cause of unprecedented radicalization of Moslems throughout the Mideast by the CIA. Young men from all over the Middle East were recruited, trained and armed as radicalized mercenaries of the United States in its "Cold War" with Russia in Afghanistan. The net result was a complete change in how the peoples of the Mideast who were predominately Moslems saw themselves and the world.

When the Afgan-Russian War ended the world had thousands of radicalized and well trained fighters on its hands who had no cause for which to fight and die.

It is easy in retrospect to challenge the wisdom of having imparted such skills to jihad-waging Islamists. These were extremists who even at the time [of the Afgan-Russian War] made it clear

---

[4] Ibid., Scott, p. 171.

they despised the West almost as much as the did the Soviet Union. But what remains is the dangerous system whereby small cliques of policy makers, acting at the highest levels of secrecy, are able to make ill-considered decisions, focused on the techniques and material of violence, that will have long-term and tragic effects worldwide. [5]

Jews and Blacks in America

Little that has happened in America since it was discovered has happened without the Jews being involved in some way. As we have seen there might never have been a United States without the Jewish input and their involvement certainly has left its imprint on us as a nation. Though, this is certainly not a Catholic nation it is a Christian one and that still irks the Jews. They have tried many times to destroy that Christian character even if it meant destroying the country. Today they are straining in the direction of upsetting society so much as to bring a dissolution of all the civilizing factors that Christianity brings.

In the United States, the Jews have long seen the blacks as their foot soldiers. What could be better? They were not just unfairly treated peasants they were actually slaves, or now the descendents of slaves. Who else would be more eager to destroy the establishment society? What they found was the blacks are simply not that type of people. That is true partly because every black alive today is several generations removed from slavery, partly because of the lavish welfare available to them and partly because they seem almost genetically unable to be part of a revolution. True, at times they can be riled up to burn down their own neighborhoods but that is in reality only something to break the monotony of their lives rather than a revolution.

NAACP

The NAACP (National Association for the Advancement of Colored People) was started in 1909 by Marxists and millionaire Jews, but no blacks. One was W. E. B. Dubois who would later become the leader of the Communist Party in the U.S. The first NAACP president was the wealthy Jew, Joel Spingarn. There was not a black president until 1966,

---

[5] Ibid., p. 118.

and the organization continues to be financed by Jews. It is one of many such groups. Why do Jews do that? It goes back to at least the Civil War. The Jews saw the American black slaves as the natural enemy of the establishment and wanted to use them to rise up in a revolution and throw off their white masters like they used the white Russians in 1917.

ACLU

The ACLU (American Civil Liberties Union) was founded in 1920 by Roger Baldwin along with William Z. Foster, and A. J. Muste all communists. Today four of nine executives are Jews and its funding comes primarily from *The American Jewish Committee*; and *The International Association of Jewish Lawyers And Jurists*. Its primary aim is to represent cases that strike at the heart of the American way of life to bring on social changes that they hope will end in making this a communist state.

Civil Rights Movement

For those of us who lived through the civil rights movement in the nineteen fifties and sixties it certainly seemed on the surface to be a black movement. However, much of the organization and nearly all the financing came from wealthy Jews. This is a perfect case of where Jews deliberately put themselves off as white. If you listen to a Jew on the matter he will say it is because both Jews and blacks are minorities and anything that helps one minority will likely help other minorities. However, Martin Luther King Jr. knew fully well that the help was not coming from whites but Jews, especially financial help.

King's top white adviser was Stanley Levison, a Jewish lawyer whom the FBI believed was a communist agent but whom King relied on to handle his finances, edit his books, and give counsel during some of the crucial crises facing the movement. The president of the NAACP and one of King's top contributors was Kivie Kaplan, a retired Boston businessman who—personally and through friends—gave hundreds of thousands of dollars, often after a hurried phone call from King or one of his lieutenants. Over at CORE [Congress Of Racial Equality], James Farmer's top fundraiser and a key speech writer was Marvin Rich, later succeeded by another Jewish civil rights advocate, Alan Gartner. Jews made up more than half the white lawyers who went south to defend the civil rights protesters. They made up half to three-quarter of the contributors to civil rights organizations, even to the more radical organizations, like SNCC [Student Nonviolent Coordinating Committee]. [6]

The book just referenced is typical of a book written by a Jew. He shows how they deliberately kept up the facade that they were nice whites from the north helping out the beleaguered blacks of the south. The claim that the Jews were involved in civil rights because they were a down trodden minority like the blacks rings hollow seeing that as a minority Jews were the most wealthy, secure, and powerful of all groups, minority or not, in America.

John Lewis who was heavily involved in SNCC was very black. He seemed less aware that his help was coming in large part from Jews.

Already these plans were taking shape, under Bob Moses's leadership. Earlier that summer Moses, [a black Harlem native] along with a white activist named Al Lowenstein, had come up with the idea of staging a "mock election" in the fall, to coincide with the actual election in Mississippi—elections from which virtually all black people were excluded. Lowenstein, who was a former dean at Stanford University, had arrived in Mississippi that summer after the murder of Medgar Evers, and he had been aroused by conditions he compared to those in South Africa. [7]

Allard (Al) K Lowenstein was a Jewish activist who had been to South Africa stirring up discontent there. He is mentioned here as one of many cases to illustrate that to blacks Jews were considered white. That is not intended to fault Lewis because most of the whites do, too. However, it must be born in mind that the blacks, no matter how well motivated, did not have the leadership skills, legal backing, or money needed

---

[6] Jonathan Kaufman, *Broken Alliance: The Turbulent Times Between Blacks and Jews in America*, Charles Scribner's Sons, 1988, pp. 85-86.

[7] John Lewis, *Walking with the Wind*, New York, Simon & Schuster, 1998, pp. 236-7.

to make the civil rights movement successful. Especially they lacked money, and many millions of dollars were spent on the movement in the south during those years. Whether or not the blacks got what they wanted out of it is unknown, but it was a perfect application of the capitalist version of the Golden Rule, namely, the man with the gold makes the rule.

This is not to say that many if not most of the Jews involved were not genuinely concerned about the plight of the blacks. This comes out clearly in Lewis' *Walking with the Wind*. However, when reading that book one sees the many ways Jews were involved. It is another example where the Jews could not have done it themselves, but without the Jews it would not have happened.

It is left to the reader to decide if the civil rights unrest of the nineteen fifties and sixties was a good thing. It is well know that the condition of the blacks was steadily improving. In fact, it might have been that advance that triggered the unrest—they wanted it to move faster. It has given us today a condition where blacks in many respects are treated as royalty. There are cases where blacks can get away with murder where a white is concerned, and if a white so much as looks crosswise at a black it can be made into a hate crime, something that will eventually be righted amidst additional violence.

Black Lives Matter

One would expect from the name that Black Lives Matter was an organization to further the interest of black people. Founded in 2013 by three black women Alicia Garza, Patrisse Cullors, and Opal Tometi it is best known for its demonstrations and obstruction of normal peoples lives through such things as shutting down freeways and airports. Such activities certainly do not make people like blacks better. But that is not their objective. Any organization that is funded by rich Jews, as this one is, whether or not it started out that way, will be maneuvered into a position of destabilizing American society. Black Lives Matter received at least $33 million in one year from George Soros, the billionaire Jew from New York, to support activists in the Ferguson, Missouri protest movement that began in August, 2014. In addition it has received $100 million in pledges from other large Jewish foundations.

Chapter 29

The Will to Power

How does one make sense out of the fact that the "theory" of evolution is the controlling authority for how people in the modern world think or as has been maintained do not think? In the strict sense we cannot make sense out of it because it is entirely nonsensical as hopefully was presented in Part I. But, it is part of our society and is not likely to go away. It means that inevitably we will be driven back to brute barbarism—a society of will.

We have mentioned the will to power several times and it is well to develop the idea with respect to the Jews and Israel. As we see in the quote within the quote below, H. G. Wells contended that there can be distinguished communities of obedience and communities of will.

The stabilizing comprehensive religions of the world, the religions which penetrate to every part of a man's being—in Egypt, India or China—gave the female principle of creation at least as much importance as the male, and would not have taken seriously a philosophy that failed to include them both. These were what H. G. Wells called communities of obedience. The aggressive, nomadic societies—what he called communities of will—Israel, Islam, the Protestant North, conceived their gods as male. It's a curious fact that the all-male religions have produced no religious imagery—in most cases have positively forbidden it. The great religious art of the world is deeply involved with the female principle. Of course, the ordinary Catholic who prayed to the Virgin was not conscious of any of this; nor was he or she interested in the really baffling theological problems represented by the doctrine of the Immaculate Conception. He simply knew that the heretics wanted to deprive him of that sweet, compassionate, approachable being who would intercede for him, as his mother might have interceded with a hard master. [1]

It is interesting to note what H. G. Wells included Israel in the communities of will. Whether or not one could say that about the Israel of the Old Testament, it certainly is true today. It could be ranked among the top communities of all time operating only on its own selfish self-interest. They have gone beyond "an eye for an eye, a tooth for a tooth" to several eyes for an eye, several teeth for a tooth.

Generally, the term "communities of obedience" is more commonly rendered as communities of reason because when a person disobeys a rule he is disciplined. From the discipline the person reasons that to avoid that unpleasantness in the future he refrains from doing the act that resulted in the discipline. Of course that presupposes the rules do not change or the whole procedure becomes pointless. The only way to keep the fickleness of man out of rule making is to base the rules on an unchangeable standard such as the Ten Commandments, that is, on the rule of God.

Opposed to the community of reason is the community of will. In this case each person does only what is in his own selfish self-interest or that of his immediate company. This inevitably leads to Darwinian survival of the fittest which is vicious and destructive. Here we are once again brought back to the beginnings of this book and can see that survival of the fittest does nothing to make new species, it, on the contrary, is a leading cause of the extinction of species, and in the human case the end of safe, peaceable societies. Or once again to quote Clark:

But no one seems to have realized how far abandonment to sensation [will] might take us, or what a questionable divinity nature might prove to be: no one except the Marquis de Sade who saw through the new god—or goddess—from the start! "Nature averse

---

[1] Kenneth Clark, *Civilization*, Harper and Row Publishers, NY, 1969, p. 177.

to crime," he said in 1792, "I tell you that nature lives and breathes by it, hungers at all her pores for bloodshed, yearns with all her heart for the furtherance of cruelty." [2]

And when we apply the survival of the fittest to economics we also enter into a community of will. "Free enterprise and the survival of the fittest: one can see how they looked like laws of nature—and in fact were both to become involved with Darwin's theories of natural selection." [3]

The community of will originates from the rejection of Logos. Here, lest it be misunderstood, Logos is meant to be the Word, the creative and sustaining spirit of God as revealed in Jesus, from the first chapter of the Gospel according to John. Having rejected Logos a society moves to a state of pointless striving or becoming with no expectation of reaching a goal; a society where the truth, the good and the beautiful do not exist.

It is the disassociation of knowledge from meaningful completion that undermines the objectivity of science and reason, and therefore the objectivity of truth. But lacking the goal of objectivity, of the notion of meaningful completion, then truth and love are replaced by fact and power. The mind and the will are at war with each other, a war that mind cannot win. [4]

Antithetical to this is the community of reason where virtue, truth and beauty can exist.

Virtue—or the exercising of responsible freedom—relies upon the existence of objective purpose or purposes, on what Aristotle refers to as *final cause* and an *unmoved mover*. It centers on the pursuit of *Being*, rather than mere *becoming*. The story of the shift from Being to becoming is the narrative of Western culture since the Baroque [1550-1750]. That deep cultural pendulum is now swinging from becoming to Being, but that shift newly faces the

---

[2] Ibid., p. 274.
[3] Ibid., p. 326.
[4] Arthur Pontynen and Rod Miller, *Western Culture At The American Crossroads*. Wilmington, Delaware, ISI Books, 2011, p. 10.

challenge of how to understand the nature and content of Being and its relationship with becoming. [5]

Of course, there is little challenge in understanding the relationship of Being to becoming if one accepts Logos as has always been intended. There is little doubt that today's problems stem largely from the relentless pursuit of becoming. No wonder nobody is happy. If we are only an endless becoming, as the evolutionists want us to believe, we must be getting better and better all the time. And, we all naturally assume getting better means becoming happier. Since we are obviously not as a society becoming happier it leaves people with a cognitive dissonance.

The Expression of the Will to Power

The will to power expresses itself in many ways but it always involves the control of others. For the Jews the process is a follows. The fanatics keep pushing the envelope forcing other Jews to either go along with their agenda or to cease being part of the Jewish community. As we have seen in today's world, a Jew being forced out of the Jewish community would be at a great disadvantage seeing as the collective Jews have control over so much of the goods and power of the world. But, this is nothing new. We see it in the case of the man born blind in St. John's Gospel, Jn 9:18-23, already mentioned.

With respect to Gentiles it is commonly seen in the form of hypocrisy and the double standard. An example of the double standard is the distinction they make between the Jews of today and the Jews of Christ's time regarding collective responsibility. Jews can hold the German people responsible for Hitler's crimes, forcing generations of German taxpayers to pay billions in reparations to Jewish organizations and the state of Israel. The whole German people is considered responsible and subsequently punished for faults committed by its leaders, even when those faults were unknown to most of the people and are now generations in the past. On the other hand, Jews zealously deny collective responsibility for the death of Christ even though the Gospel accounts make it clear that many Jewish people in Jerusalem were aware of what their leaders were doing and whole heartily supported them.

---

[5] Ibid., p. 13.

This is why evolutionists use sort of a double standard in their idea of species. It is plain for all to see that species are distinct since only members of the same species can breed and produce fertile offspring. However, if an exact definition of species is accepted it makes the concept of always becoming hard to explain. For them they want to think of all life forms as being in a process of slowly changing into something other than what they are with no sharp dividing lines between the before and after and from one species to another. That makes it easy to explain how humans came to be or rather are in the process of becoming making them nothing but smart animals. Since animals do not have reason and free will why would anyone think we do? It is all aimed at the idea of no limits in our conduct, the hedonism of "if it feels good, do it." They like to say a person can do anything he wants to do as long as it does not hurt anyone else. That last part is a lie, because going outside the bounds of morality always hurts others.

The Legal System

The will to power has perverted our legal system. One place this is seen is in that is called "hate crimes." This means that if a white person commits a crime against a black it is worse than if it is the other way around. It is assumed the white man harmed the black because he was black rather than any of the normal motives related to the seven deadly sins. From the founding of the United States our legal system was based on blind justice. That meant most of all that the law applied equally to all regardless of the social status, wealth or influence of the perpetrator or the victim. Hate crimes by their very definition turns blind justice on its head. It makes justice a function of emotion, or will, rather than reason.

This thinking even goes to the point that a person is seen more or less guilty based upon how much remorse he shows. Here again it is emotion that counts, not the seriousness of the crime. Remorse must be distinguished from the amount of rational consent that was involved on the part of the criminal. An accidental crime is different from a premeditated one.

This leads to another aspect of will over reason in our judicial system. Formerly there was a distinction made between crimes committed in hot blood versus cold blood. That distinction is still there but it is reversed.

Hot blooded crimes were those committed in the heat of emotion, normally anger, with no prior planning. Cold blooded crimes were considered worse because they were committed with malice of forethought where planning went into them. Today anger is seen as such a terrible thing because most people do nasty things to others and are afraid that if the recipient of the malice becomes angry he might harm the perpetrator. Hence, anger is worse than sneaky under-handed behavior. This has happened largely because of the great influx of women into the work force. Women are far more prone to subtly harming a person through gossip or other back handed, mean spirited means than are men. Whereas, in the case of physical violence a woman are more likely to be the loser.

Chapter 30

The United States Today

President Donald Trump

Even though Donald Trump is no longer president he was the face of America from the time of his election in November of 2016 until he was replaced by Joe Biden in January of 2021. When Trump was elected he sent shock waves through the liberal establishment like had never before happened. There are many reasons for that among them was the fact that Hillary Clinton and the left could not conceive of her losing. We also learned that the liberal establishment consisted of more than the Democrat Party and the mass media. It contained a large percentage of the Republican Party and their hangers-on in the form of never-Trumpers.

What is astounding was that the liberals thought they would have him thrown out of office within six weeks after his inauguration. They had assassinated JFK because they did not like him. They managed to force Richard Nixon to resign over what amounted to nothing, again because they did not like him. So emboldened, they expected that Donald Trump who was not a politician would be easy. What they did not count on was the fact that he was, indeed, not a politician where the old political tricks did not work.

But, what specifically is the "charm" that made it possible for Trump not to be forced out or assassinated? Certainly there would be no shortage of potential assassins. As a minor point the reason that he eats so much fast food might be that if a fast food establishment is chosen at random at a random time it would be nearly impossible to poison him.

However, that is not the real reason. To find out how he manages to survive one has only to look at the Jews.

Trump is a New Yorker. He lived there most of his life. New York is a Jewish town. He once said words to the effect that he would never let anyone but a Jew who wore a yarmulke full time handle his money. The three children of his first marriage all married Jews. Why was that? Because Jews were the people they had associated with as they grew up. As a result of all of this he in some ways thinks like a Jew and has many influential Jewish friends.

It is apparent that what Trump did was he set *his* Jews against *their* Jews. We have seen that Jews control all of the main media and that juggernaut is solidly against him as is most of the political establishment which we have also seen is either Jewish or beholding to Jews.

Look at what Trump has done. He has nullified the "treaty" with Iran that Obama signed. Israel hated that treaty. He has moved the U.S. embassy in Israel from Tel Aviv to Jerusalem something that has been left undone by numerous previous presidents and was something Israel most dearly wanted. One web site went so far as to say he was a member of the Mossad. Well, that may be a bit extreme. However, in addition to the foregoing he has put a solid block of Zionists on his staff.

In his second year in office President Trump has succeeded in putting hard-core pro-Zionists into key positions. New Secretary of State Mike Pompeo, whose first stop on his international travels was Jerusalem, is a well-known pro-Israel hawk. United Nations Ambassador Nikki Haley has become a darling of the Israeli media and will be addressing the annual CUFI [Christians United For Israel] conference in July. Trump's ambassador to Israel, David Friedman, is an orthodox Jew with decades-long ties to Israel. Friedman's views in support of Israel's expansionist policies caused five former American ambassadors to Israel to oppose his nomination in 2017. Finally, Trump has resurrected long-time neocon John Bolton, designating him as his national security advisor. [Bolton was fired in September 2019.] Bolton, who helped instigate the Bush administration's invasion of Iraq in 2003, is so overwhelmingly pro-Zionist that he had been accused of sabotaging U.S. policy in Israel's favor. With such pro-Israel officials on his staff, Trump's claim to be promoting the peace process between Israel and the Palestinians is unfortunately ludicrous. [1]

On August 27, 2018, Trump held a White House dinner for a hundred of the most influential Christians who back Zionism.

Immediately upon Trump's election the Democrats said that as soon as they got enough votes in the Horse of Representatives and the Senate they will immediately impeach Trump. How can they constitutionally do that?

The President, Vice President and all civil Officers of the United States, shall be removed from Office on Impeachment for, and Conviction of, Treason, Bribery, or other high Crimes and Misdemeanors. [2]

Ever since Trump was nominated by the Republican Party as their candidate for president the liberal FBI, the Democrats and the media began digging to find something with which to disqualify him. The Mueller Commission was formed in May of 2017 to find something that would disqualify him for being president. Robert Mueller had an unlimited budget and as many lawyers as he wanted. The main thrust was that he wanted to prove Trump colluded with Russia to taint the election and get himself elected. On April 18, 2019, Mueller came out with his long awaited report and he had absolutely nothing incriminating to report. Since they found nothing, one can reasonably assume there is nothing to find in spite of the fact that they went far afield from simply looking for Russian collusion. Notice that the Section 4 of the *U.S. Constitution* quoted above it says "high Crimes *and* Misdemeanors", not "high Crimes *or* Misdemeanors." That means they must find a serious crime that he has committed.

Or, do they? On December 18, 2019 the U.S. House of Representatives impeached President Donald Trump on two articles, both of which would not stand up in any court. The articles are as follows:

---

[1] Gary Taphorn; Israel at 70: The Blindness Continues, *Catholic Family News*, July 2018. P. 3.

[2] *U.S. Constitution*, Article II, Section 4.

Article 1: Using the powers of his high office, President Trump solicited the interference of a foreign government, Ukraine, in the 2020 United States Presidential election.

Article 2: President Trump abused the powers of his high office through the following means:
(1) Directing the White House to defy a lawful subpoena by withholding the production of documents sought therein by the Committees.
(2) Directing other Executive Branch agencies and offices to defy lawful subpoenas and withhold the production of documents and records from the Committees—in response to which the Department of State, Office of Management and Budget, Department of Energy, and Department of Defense refused to produce a single document or record.

Notice in Article 1 that having found nothing where he affected the election of 2016 they now resorted to the 2020 election. The issue with Ukraine was common diplomacy. Trump early on found that the career diplomats from the State Department leaked everything he did so he cut them out of the loop. This article is in retaliation for that snub.

Article 2 is in retaliation for another snub where he did not cooperate in turning over documents where he had no requirement to do so.

As expected the Senate which had a slim Republican majority did not convict President Trump just as a slim majority of Democrats did not convict President Bill Clinton.

Surprisingly they did not resort to the XXV Amendment to the *Constitution*. It has to do with the succession of power if the president dies or is incapacitated. It was ratified February 10, 1967, which was a little more than three years after the assassination of President Kennedy. It is likely Kennedy's death was the reason for it even though there was no problem with the succession of power since within hours of the assassination, Vice President Johnson was sworn in as president.

Here we see once again the Jews and Israel at work. The XXV Amendment deals primarily with the case where the president is incapacitated something that the original Constitution did not address. While the nation was in shock at what happened to JFK those in power behind the scenes saw the chance to push though an amendment that the founding fathers never intended; something for the future in case an undesirable man became president. And, Donald Trump is such a man.

The salient part of the twenty-fifth Admendment is Section 4.

> **Section 4**. Whenever the Vice Presiden and a majority of either the principal officers of the executive departments, **or of such other body as Congress may by law provide**, transmit to the President pro tempore of the Senate and the Speaker of the House of Representatives their written declaration that the President is unable to discharge the powers and duties of his office, the Vice President shall immediately assume the powers and duties of the office as Acting President. Emphasis added.

The clause "or of such other body as Congress may by law provide" is completely without constraint; it could mean anything. From the way it is written it is not clear that such a body must include the Vice President. It could be only two people. And, do not worry about the word "law." The Mueller Commission was illegally started. There must *by law* be evidence of a crime for such a commission to be started and there was no such thing.

What people were thinking about when they ratified this amendment was the case when the president has taken ill, for example, by a heart attack. But, from the start the liberals have been saying that Trump was not suited to be president. There was nothing specific other then they did not like his style or that they did not like him because he was an outsider.

More than half of the electorate thought Donald Trump was a better choice than Hillary Clinton and that level of support has not shown signs of weakening.

What Did Donald Trump Do Wrong

If Trump was so popular why did he have so much trouble dealing with congress and finally getting defeated for a second term? The answer to the first question is that he came from the world of finance, entertainment and construction. In those fields when a person goes into negotiations he assumes that both sides are after the same thing, namely money.

But, in politics, especially U.S. politics, that is not so. Both sides are not after the same thing. Conservatives want to make a strong county with a thriving economy while liberals want to tear America down. They wouldn't say it that way but that is what all of their policies are aimed at.

As far as not getting reelected goes, he wasn't listening to Barack Obama after his first win. It is all but impossible to pin this down years later but at the time he made statements to the effect that he would quietly go about the country setting up cells in every city and town that would be organized to act in unison when the time came for some unspecified action. As this was happening this writer remembers getting a couple of phone calls that subtly asked if there was interest in this or that cause. In hindsight we now know that what Obama did was set up cadres in each voting precinct so there would be a preponderance of left leaning people available to help with the election from pole watchers to vote counters. In the 2020 election my wife and I saw our ballots negated by just such people.

For the 2016 election Trump took his message directly to the people and he won. At the time that was about all he could do. But, that would never work a second time. He needed the inside, down and dirty organization just like the Democrats had. It is reminiscent of how Pope John Paul II decided not to fight the ingrown bureaucracy of the Catholic Church and took his message directly to the people by traveling all over the world and preaching. In the end it amounted to nothing because he did not get a handle on the real source of the problems in the Church.

America Divided

It might be well for the reader to go back and revisit the quotes from Whittaker Chambers. In 1923 he traveled to Europe and saw the devastation from World War I. With the coming of the Great Depression after the stock market crash of 1929 in the United states and, indeed, most of the world, communism took hold outside Russia. Chambers was a member of the Communist Party in America through most of the depression. During that period the U.S. came dangerously close to going communist. It was the Second World War that put that progress on hold. However, it never went away.

Today we are a nation divided by the idea of national socialism which links to International Communism as opposed to a free market republic. The election of Donald Trump in 2016 brought our division into sharp focus. Those who voted for Trump are members of the society of reason. Those who voted for Hillary Clinton are from the society of will. Granted, members of the two groups, in various ways and times, drift to the other side, but in the main there is no doubt where the loyalties of individuals lie.

In the summer of 2018 President Trump nominated Brett Kavanaugh to replace retiring Supreme Court Justice Anthony Kennedy. The left went wild where there was no reason to do that. Ostensibly it was because they were deathly afraid the court will be packed with a pro-life majority and the abortion question might be revisited. From Kavanaugh's record he was not particularly pro-life. That did not matter. If Trump nominated him, he had to be bad. In his defense it must be noted that he did side with the majority when they over turned the Roe vs. Wade decision of 1973.

Some time after Kavanaugh's nomination a woman, Christine Blasey Ford, surfaced with the accusation that he sexually assaulted her at a house party when she was fifteen and he was seventeen. Finally on September 27, 2018, after weeks of delay and obfuscation on the part of the Democrats she finally testified before the Judiciary Committee. There were few facts other than the accusation. She claimed to have been assaulted, not raped, by Kavanaugh and another young man she called Mark. She did not remember where the house was, nor the month nor the day. The Mark turned out to be Mark Judge who when interviewed by the FBI knew nothing of the party or incident in question. Amazingly through all of her dim memories the only thing she remembered with absolute clarity was Brett Kavanaugh.

When all the statements she and others made were carefully analyzed it turned out that Ford's testimony was all lies; that she had come forward with the expectation that the mere accusation would cause Kavanaugh to withdraw. That did not matter to the Democrats. It got so bad that the four senators who were undecided about Kavanaugh each had to be accompanied by three armed policemen just to walk the halls of the Senate Office Building. The building was swarming with protestors. Even the offices of those senators were packed with protestors. How that was allowed in a place like that is a mystery to normal people. But, with

Jews making the rules about who gets in, what else would one expect.

All of the delays and lying about Kavanaugh were about abortion as everyone knew but none dared say. A Republican senator from Maine, Susan Collins, who was one of the hold-outs let that be known for certain. She relayed in the senate chambers that she decided to vote for Kavanaugh because he personally told her that he considered abortion to be settled law and he would do nothing to change that. On Saturday October 6, 2018 starting at five o'clock in the afternoon the final vote was taken and he was confirmed 50 to 48.

By the time this work is published the Kavanaugh incident will likely be lost to the collective memory. The reason it is mentioned is it brings into sharp focus the contrast between the conservatives and liberals. For conservatives, which we are calling the society of reason, an accusation must be backed up by witnesses or evidence. This is a rule going back at least four thousand years in societies of reason, as for example the ancient Egyptians, Indians and Chinese. The obvious reason for the rule is because people lie. With no witnesses or evidence nothing would have been served by that woman coming forward with her claims.

But, for the other side it made perfect sense. For them the facts of the case were irrelevant. What mattered was the accusation and her sincerity. That is, for the society of will, sensitivity and emotions are everything. To mention just a couple of comments by the liberal elite in the media concerning Ford's testimony, Brit Hume of Fox News said, "The more hesitant, the more fragile she has seemed, the more credible and powerful she seemed to the audience." That is, hesitance and fragility easily make up for there being no witnesses or evidence. NBC called her testimony "gut wrenching." This was a case of what is now called the "me too" movement. That means that a woman can accuse a man of misconduct with regard to her and that means it is true. The man will categorically be assumed to be lying which does not bode well for our future.

On a more general level the vitriolic discourse one hears in public by the liberals has reached a level formerly unknown. No matter that most of the acrimonious voices are either in the employ of media outlets or are paid demonstrators the sheer hate expressed produces a coarseness in society. Even more disturbing is the fact that speech in itself has reached its limit of effectiveness which means that more and more physical violence will be seen as necessary.

This is why maintaining the doctrine of evolution is so important to the left. When we start with animals who only live in a world of will it is easier to say people are no different. As we have seen, it is technically wrong to connect animals with will because they operate solely by instinct and do not have free will. They are the ultimate case of "me first" because that is the way they are programmed. They do not operate on the bases of the survival of the fittest but on the survival of "me," and that is how liberals operate.

Our present divide into Democrat and Republican, left and right, liberal and conservative, will and reason got a big push with the civil rights movement. "American society today is divided by party and by ideology in a way it has perhaps not been since the Civil War. . . . I can end part of the suspense right now—Democrats are the winners. Their party won the 1960s—they gained money, power, and prestige. The GOP is the party of the people who lost those things." [3] Notice that Caldwell does not mention that rich Jews are the ones with the money and power and there could be legitimate disagreement as to whether the Democrats ever had prestige.

Caldwell's contention is that it was the civil rights legislation of 1964 and 1965 that gave unprecedented new powers to the federal government. ". . .it gave Washington tools it had never before had in peacetime. It created new crimes, outlawing discrimination in almost every walk of public and private life." [4] It would not have had an effect that has lasted to this day if, as intended, the laws had been confined to the South and scaled back as integration took place. But, they were extended to include gender, immigrants, homosexuals and other fringe groups. As sexual morality degenerated in the sixties it led to the Supreme Court deciding in 1973 that abortion on demand could be found in the Constitution. It was a short step from there to making homosexuality a "civil right" and all the other perversions that have followed.

These policies, qua polices, have their defenders and their de-

---

[3] Christopher Caldwell, "The Roots of Our Partisan Divide," p. 1, *Imprimis*, Vol. 49, No. 2, Hillsdale College, Hillsdale, MI, February 2020. This is from a talk summarizing his new book *The Age of Entitlement: America Since the Sixties*.
[4] Ibid., p. 2.

tractors. The important thing for our purposes is how they were established and enforced. More and more areas of American life have been withdrawn from voters' democratic control and delivered up to the bureaucratic and judicial emergency mechanisms of civil rights law. Civil rights law has become a second constitution, with powers that can be used to override the Constitution of 1787. [5]

What that means is the United States is irrevocably slipping into communism. After Trump's election the left has shed all pretenses of living by the Constitution and that its goal is a totalitarian state.

A few words should be devoted here about why the left hates Donald Trump with such a burning passion and it is well to remember that it is the Jews who drive that passion. As has been mentioned, Hitler opposed the Jews, at least at first, because they were communists. He used the idea of a pure Arian race to sell the German population on the idea that they had to stick together as a race or be defeated. Today the Jews see Trump as another Hitler and his campaign slogan of "Make America Great Again" they see as counterpoised to Hitler's pure Arian race. They see Trump as one who would call them out and drive them out for the corruption they are permeating into American society just as they have done in their host societies at various times in the past. And worst of all, at least subconsciously, they know they would have it coming. But, there is a difference today. In the past the host societies drove them out and they settled in other regions. That wouldn't work for the most part at present because of the way the world is interconnected. In so many ways they control all of human society so they could continue to exert their influence from any where.

The First American Revolution

We are facing the real possibility of a revolution in America. No, not the second revolution because the war that resulted in the formation of the United States was not a revolution. It was, as we have said, the *War of Independence*. The difference is not semantic. A revolution is where the existing structures of society—government, religion, economic practices, etc.—are over thrown. In the case of America, each of the colonies

---

[5] Ibid., p.5.

already had a government and those other practices that made society work. The war was primarily about throwing off the taxes of England which had become little more than an absentee landlord. There was no attempt to overthrow the government in England. It should be noted in passing that the difference between what is called a civil war and a revolution is a civil war is where those who start it lose and a revolution is where those who start it win.

This will be the revolution that has been gestating for decades and the election of Donald Trump as president seemed to have brought to light the sinister plotting of the left. The more Trump succeeded the more likely was a genuine revolution. When will it start and what will it look like? As for the start all revolutions need a spark. Vladimir Lenin's newspaper was called *Iskra*, "The Spark." The spark can be anything and not necessarily something significant or instantaneous. In 1917 Russia the spark was the devastation brought by the First World War. Today one gets the feeling that the left is constantly trying to generate the spark. The relatively minor incident in Ferguson, Missouri, August, 2014, through the lavish infusion of cash by revolutionary Jew, George Soros, could have been one. It failed to catch fire though the smoldering effects are still being felt as police, whose fingers were burned in Ferguson, refuse to maintain order in the slums of cities like Chicago and Baltimore.

Covid-19: *Iskra*, "The Spark"

It could well be the Covid-19 (from CoronaVirus Disease 2019) is the spark the left has been looking for. It was hardly a footnote to the news prior to March 2020, though it now appears certain to have originated in Wuhan, China the last months of 2019. It is known that as soon as it appeared, China quarantined Wuhan from the rest of China, but continued to allow air travel from Wuhan to the rest of the world thus making it a world pandemic. There are rumors that it was genetically engineered to attack old people because of the exploding number of retirees all over the planet. For a time if a person called a doctor or dentist he found that some were not making appointments for anyone over sixty-five—in you face age discrimination. That lent some credence to the age rumor.

Starting the last full week of March many states issued a two week "Stay at Home" order. All "nonessential" business were closed. This included all restaurants, pubs and most other businesses. All churches were closed and all public gatherings were prohibited. Of course, it was surprising what was considered essential. All abortion clinics and liquor stores were left open. There in an instant went two of the ten original Bill of Rights, namely freedom of religion and freedom of assembly. The worrisome thing about this was how totally the population accepted these orders, and the religious leaders did not say a word. In two months the three years of prosperity that President Trump had brought to the country disappeared. The U.S. went from full employment were anyone who wanted a job had one to thirty million unemployed—the worst since the great depression in the 1930s.

About the time the first two week stay at home order was to expire it was extended another two weeks, and then additional two week extensions kept coming. Most of the "Stay at Home" states were Democrat states. Some Republican states never did close anything. Others opened their states after four or five weeks with no ill effects giving the lie to it ever having been necessary.

As of the middle of May, 2020, there had been 86,000 reported deaths from Covid-19. These numbers were artificially inflated as many non-Covid deaths were chalked up to it to keep the numbers. Someone dying of lung cancer was automatically marked up as a Covid causality. Added to that was the fact that about 80% of deaths were of people in assisted living facilities. To put that in prospective, most flu seasons the U.S. have between 50,000 and 60,000 flu deaths. Also, in 1957-58 we had what was called the Asian Flu which caused 70,000 deaths in the U.S. That was when the population was half of what it is now and there were far fewer old people so adjusted to today's demographics the Asian Flu would have produced upwards to 200,000 deaths. Yet the Asian Flu was taken in stride as nothing but a somewhat more severe flu season.

Shortages showed up in supermarkets. There were a lot of items that were not to be had. Hog farmers killed their own pigs because the packing plants were closed. Farmers in California plowed under tomato fields because processing plants were closed. Communists do not see growing, processing and distributing food as essential services. One only has to look at Russia where the people lived on the edge of starvation for seventy-three years to prove that.

It was so political that even all schools, colleges and universities were closed from the middle of March for the rest of the year. And the government schools K through 12 are the life blood of the liberal mantra. Along that line even professional sports took a hit. The NFL football season of 2020 had no fans in the stadiums.

As time went on the authorities, read Governors, kept making things harder. A month into the shut down a person could go into any store without a mask to cover his nose and mouth. Then suddenly they were necessary in all stores. There was murmuring that they will not back off until there was "certainty" that not another person would die from Covid-19. That shows how unhinged the left had become.

By January 2021 when Biden took office as president the nation and the world were hit with the COVID shots, better known as the Fauci-ouchie after Dr. Anthony Fauci (Dr. Jab) who was director of the National Institute of Allergy and Infectious Diseases during the Covid pandemic. He was responsible for the mask mandates and lock-downs that were later shown to be of no use. These shots were unprecedented in that they were never needed, didn't work and were very dangerous. The normal testing and approval processes for a new vaccine were thrown out with an Emergency Use Authorization. The shots later acquired the appellation of the clot-shot because they caused blood clots.

The shots were and are dangerous and do not attack only old people. "From February 2021 to March 2022, millennials [born between 1981 and 1996] experienced the equivalent of the Vietnam War, with more than 60,000 excess deaths. The Vietnam War took 12 years to kill the same number of healthy young people you've just seen die in 12 months." [6] It is known that between three and four million Americans have died from the shot and many more millions disabled from months to indefinitely. In addition Dr. Anthony Fauci (Dr. Jab) stated in 1999 that these experimental genetic shots could have effects that only manifest themselves after ten to twelve years. If that weren't bad enough data shows that not only don't the shots prevent Covid but actually make the patient more susceptible to it with each additional booster.

As this revolution unfolds it will not be like revolutions in the past. More likely it will be like in China where the government uses sophisti-

---

[6] Edward Dowd, *"Cause Unknown": the Epidemic of Sudden Deaths in 2021 and 2022*, NY, Skyhorse Publishing, 2022.

cated surveillance tools to keep records on everything an individual citizen does. They use what is called the Social Credit Score. That is a score of how politically correct a citizen is. The higher the score, the more preferential treatment one gets. Nothing is left uncontrolled or unknown. People who do unapproved things disappear. What is known about China is starting to be done in America but is unknown to the public. We are living through the communist revolution in America.

Health care, as noted above, will be an obvious venue for the will to power to operate. Having debased themselves by either performing abortions or turning a blind eye to those who do, many healthcare personnel will not flinch at deciding who gets life saving care and who dies. Years ago a nurse in the end of life care business said the over half of the people who died had some form of active intervention by doctors to speed things along. That is to say nothing of the doctors who simply do nothing to prolong life. With the massive growth in people over sixty-five as the baby boomers retire it will be seen as a social imperative to thin the ranks of the aged. Who will get thinned? The surveillance state will have the data to make those decisions.

The Growing Presence of Evil

The legislation in the 1960s aimed at helping along integration in the south was accompanied by other factors operating in western society. As one looks at the Second Vatican Council from 1962 to 1965 one sees it brought on the Catholic Church the disease worse than Covid-19, it was modernism. It is frequently said that the falling apart of secular society also infiltrated the Catholic Church when it was really the opposite. After the Council fifty-five to sixty-five percent of all Catholics left the church. The falloff of Mass attendance resulted in a rend in the flow of God's graces to the world.

> How did we get here? . . .We can be sure of one thing: all grace comes into the world through the Catholic Church: thus a dying world is the result of a struggling Church. The health of the church is reflected in the health of the world. . . . If something happens to the life to the Church, something happens to the life of the world. We need not look any further than the fateful decade of the 1960s.

. . .It is not a coincidence that the decade during which the Second Vatican Council took place was also the decide of supercharged moral and societal degeneration in a formerly Christian world. Contraception, recreational drug use, promiscuity, demonic music, abortion, New Age spirituality, divorce and Godless philosophy were established through courts and other means as pillars of the modern world. There were of course inklings of this sort of thing in the years leading up to that fateful decade, but the 60s era stands out as a time when men began to "call evil good, and good evil: that put darkness for light, and light for darkness: that put bitter for sweet and sweet for bitter" (Isa. 5:20). [7]

On February 24, 2022 Russia invaded Ukraine. It seems they thought it would take little more than a week before they had complete control of the country. It hasn't turned out that way. The Ukrainians fought back with greater skill and determination than the Russians had anticipated. In addition it wasn't long before the NATO nations, especially the U.S., stepped up the supply of arms and advisors to Ukraine to the point of seriously depleting our stock pile of certain items such as antiaircraft missiles.

If war isn't bad enough there is a particularly troubling dark side to this one. On 11/2/22 Tucker Carlson of Fox News repeated the following statement made by Jamie Raskin Democrat Congressman of Bethesda, MD on 10/28/22.

"Russia is an Orthodox country with traditional social values. That is why it must be destroyed, whatever the price paid by us. The war in Ukraine is not a conventional war but a jihad." Jamie Raskin said that out loud but many in Washington agree with him. [8]

Notice it is because Russia is still at least nominally a Catholic country with traditional social values that it must be destroyed. And it is horrifying to see how bellicose has become the invective of the enemies of

---

[7] Kennedy Hall, "Archbishop Lefebvre, Model of Spiritual Fatherhood", *Catholic Family News*, April, 2020, p. 5.

[8] https://www.newsweek.com/fact-check-did-jamie-raskin-call-destroy-christian-russia-jihad-1756056

reason.

There are now many voices saying that the war in Ukraine is the start of WW III. The other world wars started as regional conflicts so that could come to be. In this regard China is a growing threat to world peace as it has engaged in what it calls total warfare against free countries and has stated publicly that the United States is its principle enemy. Total warfare aims at harming the U.S. by subverting our whole society including education, defense, elections, news outlets, healthcare, finance, industry and anything else that is seen as vulnerable. That is, they use our open society to do damage to it, and we just let them do it. This war may in a short time erupt into armed conflict over Taiwan. China has recently increased its rhetoric and provocative acts toward Taiwan demanding that it has a right to control Taiwan even though the Chinese Communist Party, the CCP, that despotically rules mainland China has never ruled Taiwan.

There have been recent articles reasoning that it would never come to a shooting war with China over that small island. Among the many reasons is China is so totally dependent on the rest of the world, especially the U.S., for trade; crossing the Taiwan Strait and landing an invasion force would be extremely perilous and many other logical reasons. On the other side Xi Jinping is running out of time since the one child family that was in effect for several decades is having a devastating effect on the demographics of China—there are an ever decreasing supply of young men for soldiers. Another concern is that Jinping is a megalomaniac and as such could rashly attack Taiwan expecting a quick win just as Vladimir Putin did in Ukraine. Worse than that is the prospect that the use of tactical nuclear weapons are increasingly seen as a viable option in any discussion of the next major war.

A case where the U.S. Federal Government has over reached its bounds was revealed by a whistle-blower in the FBI who made public a memorandum saying that traditional Catholics were suspect of being subversives and warranted being watched. The memorandum was from the FBI, Richmond, Virginia office some of which follows.

FBI Richmond, VA, 23 January 2023

Executive Summary

FBI Richmond assesses the increasingly observed interest of racially and ethnically motivated violent extremists (RMVEs) in radical-traditionalist Catholic (RTC) ideology almost certainly presents opportunities for threat mitigation through the exploration of new avenues for tripwire and source development. FBI Richmond makes this assessment with high confidence based on FBI investigations, local law enforcement agency reporting, and liaison reporting, with varying degrees of corroboration and access.

In making this assessment, FBI Richmond relied on the key assumption that RMVEs will continue to find RTC ideology attractive and will continue to attempt to connect with RTC adherents, both virtually via social media and in-person at places of worship. [9]

What that says if someone attends a traditionalist Mass he could expect to have FBI agents there taking names. If they are to see people who meet "in-person at places of worship" the agents must be there themselves. This is another case where the Department of Justice and the FBI in particular has been weaponized against law abiding conservative citizens.

In a world that decries Christian spirituality it may not seem odd to say that as the grace of God decreases and dies evil increases. We are seeing the spread of evil over the globe now as never before. Whether or not additional devils are released from hell to prowl the earth to torment men as each Catholic Church is closed or those already here are driven to greater zeal for darkness is not clear. When a large percentage of society gives in to its appetites and passions, when satanic worship is on the rise it can only mean that the devils are taking charge with all the horrors that entails. Humanity is staggering under the weight of sin. The light of God's grace that had been a stabilizing factor on Western Civilization since the Middle Ages has been withdrawn and darkness is covering the earth.

---

[9] https://www.uncoverdc.com/2023/02/08/the-fbi-doubles-down-on-christians-and-white-supremacy-in-2023

Chapter 31

What Lies Ahead

Though the citizenry of the United States of America through its history have as a whole had a strong sense of right and wrong, we have never had a unifying philosophy other than that of profit with Christianity in the form of Protestantism being the prevailing religion. As we have seen Protestantism leans heavily toward predestination and worldly wealth. The vast majority of people who came to our shores did so for a better life. Freedom was important, but it was the freedom to keep the benefits of one's labors that mattered most. To a lesser degree it was the freedom to worship according to one's own beliefs. Certainly since the First World War, the profit motive has been transcendent. Opposing us were the Jews who had always been there gnawing away at society. But when we lost the organizing influence of Christianity, the Jews who are always organized found it easy to take over. And, they have no intention of relinquishing control.

The other way to look at how it happens that Jews control the U.S. is to say they have always done so. This goes back to what has been said about the Masons having a controlling influence on our nation from its founding. It has been said that there has never been a president of the Unites States who was not a Mason. That statement is not officially acknowledged, but many of them were.

An acquaintance of the author once mentioned that he thought it was obvious that Protestantism was superior to Catholicism because the U.S., being Protestant is so prosperous, and all of Central and South America being Catholic was so impoverished. On the surface that is true vis-à-vis prosperity and poverty but it in no way explains the root cause of the disparity which it is not religion per se. It is caused by the Masons, which is to say the Jews. There have been thriving economies south of the border many times. But, outsiders have always come in and destroyed them in what is the continuing war between the Jews and Catholics.

What happened, in fact is still happening, in Central and South America is a slow motion version of what happened in Russia where the results were a stark, almost instantaneous change.

The Jewish End Game

As we have noted, at the time of Christ there were already Jews living in all parts of the Roman Empire. In 70 AD the Romans had finally had enough of the Jews in Palestine and utterly destroyed the Temple and Jerusalem. Those Jews not killed were driven out to join the Diaspora communities in Arabia, Egypt and other nations. Through the centuries those nations, countries, states, principalities and fiefdoms that had a strong moral and religious character drove out the Jews. As we have seen in 1492 the Spanish under Queen Isabella and her husband Ferdinand finally drove out the last of the Moors (Moslems). They also sent the Jews packing since they were as big a menace as the Moors, though in different ways. Many of the Jews thus expelled migrated to Germany, Poland and Russia. That is the main reason there were as many Jews in those lands as there were at the start of the twentieth century.

The reason the above instance is mentioned is that it might have worked to drive Jews out of your country those centuries ago. But, it would not do much good today because the world has become so interconnected. It is doubtful that there are many Jews in China. But, if the world Jewish community wanted to crash China's economy they could do it because China needs access to the rest of the world for raw materials and markets for their products. It is worth noting that the U.S. Fortune 500 companies make up about two-thirds of the U.S. GDP and they are all heavily invested in China. And a preponderance of the control of these companies is in the hands of Jews. If it happens that China becomes a poor investment and not just the Fortune 500 companies but the myriad other companies pull out, China will suffer.

The point of all of this is that if today Jews should come under attack anywhere in the world the Jewish community could immediately retaliate since they control so much of the world's finances and major industries.

The 9/11 attacks were extreme and if it ever became widely known that Israel was behind them, things could turn ugly for Israel. Realizing that, they would have taken steps to keep America in line. Consider this. Israel is thought to have somewhere between 300 and 400 atomic bombs. What is to stop them from planting one in Manhattan Island, one in London, and for that matter in many of the major cities of the world? It could be done and the devices could lay in wait for decades with none of the "host" cities being the wiser. They could be set up so it would only take a code from an international phone call to detonate them. In fact, some say that has already been done. "We now know that Israel has planted 25 nukes inside America in major Cities as a part of their infamous 'Samson Option.' Some believe they even planted more in major European Cities, such as London, Paris and Rome". [1]

It is important in this regard to distinguish between atomic bombs and hydrogen bombs. Hydrogen bombs are by far the most powerful, though even an atomic bomb the size of the one dropped on Hiroshima would easily wipe out New York or London for all practical purposes. Hydrogen bombs use an unstable isotope of hydrogen called tritium. Tritium has a short half-life of 12.5 years. This means that in 12.5 years half of it is gone so all of the hydrogen bombs setting on missiles in silos and elsewhere have to have their tritium replenished every few years. That is not the case with atomic bombs. Atomic bombs use either plutonium Pu-239 or uranium U-235. Pu-239 has a half-life of 24,000 years, U-235 has a half-life of 700 million years. That means from the point of view of a century of two they do not age at all.

Ultimately, the Jews—here we mean the Talmudic Jews—mean to totally control the world. Since they believe only they are truly human and as such have rights to all of the world's wealth, in the end they simply want it all. That is their goal, that is their end game.

Can the Jews Achieve Their Goal?

The premise of this book is that we have let Jews control what we

---

[1] https://www.veteranstodayarchives.com/2014/10/07/preventing-the-transformation-of-america-into-gaza-ii/

may think and for most of us have kept us from thinking at all. Worst of all we, through a glass darkly, know this is the case. Our mental processes exist in a corporate void. Without thinking on the part of the citizens a totalitarian world awaits us as many uninformed people vote for socialist candidates. And do not try of change any minds. Once again it is like those who mindlessly embrace evolution. "From contemporary experience with totalitarian movements it is well known that the device [socialist propaganda] is fairly fool proof because it can reckon with the voluntary censorship of the adherents; the faithful member of a movement will not touch literature that is apt to argue against, or show disrespect for, his cherished beliefs." [2] There are still people who think but even those are guarded as to whether or not they may even have certain politically incorrect words in their thoughts to say nothing of speaking or writing them.

This author recently heard a song from the sixties from the ubiquitous overhead speakers of a fast food restaurant. The song was the Simon and Garfunkel hit song *The Sound Of Silence*. The lyrics of the third stanza go like this:

> And in the naked light I saw
> Ten thousand people, maybe more
> People talking without speaking
> People hearing without listening
> People writing songs that voices never share
> No one dare
> Disturb the sound of silence.
> © Columbia Records

It is doubtful that Paul Simon who wrote the song was a philosopher. However it does tell us a lot about ourselves. The title is a humorous play on words. The lyrics, on the other hand, are pessimistic, almost depressing. Considering it was first released in 1964 it was, if nothing else, prophetic of today's world.

The lines "People talking without speaking, People hearing without listening" strongly remind us of the all too common phenomenon where

---

[2] Ibid., Voegelin, p. 140.

two people are setting across the table from one another and both are completely engrossed with their smart phones to the exclusion of all others including their table companion. Maybe the reason smart phones have taken over the world so totally is because it frees the user from having to communicate with other people in person and face the prospect of slipping and saying something to which the thought police would object. Of course it could be a play on the words of Jesus in the gospels. For example in Matthew 13:13 he says, "This is why I speak to them in parables, because seeing they do not see, and hearing they do not hear, neither do they understand."

Another possibility is that the "sound of silence" does not refer to physical sound because we live in a noisy world. What made such an impression on me in the restaurant was the fact that the song was barely detectable over the background noise—the drink machine dumping ice into a cup, the server calling out the number of the order that was ready, people coming and going—and what was the song? *The Sound Of Silence.*

It is not physical sound that is silent, but spiritual sound. On the most basic level, all thought is spiritual because it occurs in our spiritual soul that is in a mysterious way connected to our physical brain.

There is no denying we have let ourselves come to be in mental bondage by the politically correct police—the Jews—and short of a direct act of God, it is not likely to change. Is a direct act of God likely? Probably not but not impossible. It has happened before in modern history. The reference here is to the conversion of Mexico to Catholicism after the appearance of Our Lady Of Guadalupe in 1531. The Spanish priests had been making slow progress against the native religion that demanded thousands of humans be sacrificed each year. In the seventeen years after Our Lady's appearance nine million Mexican Indians were converted and baptized thereby eradicating the pagan religion. It is the most sudden and total conversion of an entire people ever recorded, proving that it can happen. And through the intervening centuries the pagan religion never came back.

Catholic Indian communities were created throughout Mexico. In spite of epidemics and later oppression, in spite of the Church's own mistake of failing until recent times to ordain native priests, in spite or revolutions and bitter anti-Catholic governments, the

Catholic Faith has endured in Mexico, just as the beautiful portrait of the Mother of God still glows undimmed from Juan Diego's tilma in Mexico City. [3]

This can be juxtaposed to the Israelites of the Old Testament. They fell into paganism and had to be reconverted to their true religion dozens of times. So, would something like this ever happen to the Jews today? The prophecies of the end times say it will but a lot can happen short of the crack of doom.

Short of divine intervention there is always that thing that happens to a specific people when they come under total control of a corrupt elite. The adage that power corrupts and absolute power corrupts absolutely comes to mind. When some person or group turns to evil and acquires control over a society they unleash the evil they have embraced upon the citizens beneath them forming a tyrannical regime.

Opposed to this is some thoughtful Jews are beginning to express doubts that the extreme Zionism that has been driving the whole Jewish community since 1948 is not producing the desired results. It must be stressful living in Israel, the most despised nation on earth, where all your neighbors have as their number one priority you destruction. That is, is there some third way between being the bullies of the world and the threat of being eradicated that would be a more direct route to their desired heaven on earth? The answer to that is probably no because neither they nor anyone else will ever establish paradise on earth.

Another way to look at it is to consider the Gentiles. A religiously strong and morally sound society of reason will force out Jews as happened in England in 1290, the French in 1306 and Spain in 1492 as mentioned above. Opposed to that a society of will, which if brutal enough, as in the case of the Moslems, will do the same thing. Since most of the Jews around the world live in what traditionally were societies of reason, that would mean the peoples of western Gentile countries would have to once again become strong, devout Catholics.

The West today is a neo-pagan society offering human sacrifice in the form of abortion and euthanasia. So, like it was with the Mexicans, we could have a total conversion too. Yes, it could happen that way but that

---

[3] Ibid., Carroll, p. 103.

is not likely for two reasons. The first is that having once been a moral society based on the teaching of Christ we fell from a lofty height. This will cause us to be more intractable when it comes to conversion. The second is that we are not trying to appease the gods of wrath as did primitive peoples. We are committing murder plain and simply, and we fully well know it. As such, God will be severe in exacting a penalty.

So, will the Jews ever achieve the goal of total domination? No. There are two antidotes that are appropriate in this regard. The first is when Napoleon told a priest that he intended to totally wipe out the Catholic Church, the priest replied that the Catholic clergy had been trying to do that for eighteen hundred years. If they failed why did he, one man, hope to succeed? The other was made by an Anglican historian. Concerning the Catholic Church he quipped that for any organization so poorly managed and ineptly lead to survive for two thousand years it had to have divine assistance.

Armageddon

It will not be proposed here that we are approaching the end times. However, short of the final conflict a lot can happen. In his book, *The Book Of Destiny* [4], Fr. Kramer proceeds through the *Book Of Revelation* verse by verse and applies an interpretation to it. This has been done by many others but his work had received a lot of respect in that it is frequently quoted. He aptly shows how chapter 6:12 applies to the Arian heresy in the third century, how chapter 8:8-9 names Islam, 8:10-12 applies to the Great Schism of 1054, and how chapter 9:1 describes the Protestant Revolt. In Chapter 9:18 it says a third of the earth's population will be destroyed. It must be kept in mind that there can be considerable time between events as described. Following the annihilation of one-third of the earth's population the Beast is released from the Bottomless Pit. There could be centuries between the two events.

The future is, of course, indeterminate and the direct of hand of God in history is possible as illustrated above, but there is one strong possibility that is discernable if one looks carefully at what is happening with artificial intelligence today.

---

[4] Rev. Herman Bernard Kramer, The Book Of Destiny, Belleville, Illinois, Buechler Publishing Co. 1955,

Artificial Intelligence

We have come to a point in our history where artificial intelligence (AI) must be taken seriously. Nearly everyone knows how useful something like a spelling checker in a word processor program is to say nothing of the word processor itself. Few realize just how big a step that was going from nothing to the first desk computers where the entire operating system was on a single 5-1/4 inch floppy disk. A second disk contained the word processor, the spelling checker and the entire dictionary, while a third disk held your documents. Today a desk computer is millions of times faster and as many times more capable. And that is only the bush league. AI is already at a point where it can solve problems that no human can handle. Why would we think all of this is only for our benefit? If history tells us anything it says new technology will work both ways. In 1866 Alfred Nobel invented dynamite and there were those who said it meant the end to war because dynamite was simply too destructive. Sorry to say they were wrong.

Presently we have with what is called narrow-domain AI or artificial narrow intelligence, ANI. That is where sophisticated computer programs do very specific things like translating languages.

Already, we have trouble understanding our narrow-domain AI. In 2017, *Digital Journal* reported: "An artificial intelligence system being developed at Facebook has created its own language. It developed a system of coded words to make communication more efficient. Researchers shut it down when they realized the AI was no longer using English." Google's translation AI seems to have done something similar, and it wasn't shut down. In 2016, *New Scientist* reported on the breakthrough advance: "Google's researchers think their system achieves this breakthrough by finding a common ground whereby sentences with the same meaning are represented in similar ways regardless of language—which they say is an example of an 'interlingua.'" "In a sense, that means it has created a new common language, albeit one that's specific to the task of translation and not readable or usable for humans." [5]

---

[5] *New American*, The Age OF Artificial Intelligence Is Here, May 20, 2019. p. 18-19.

The above quote concerns domain-specific AI. What happens when several and then many domain-specific AI programs become linked? "Soon, perhaps very quickly, such an AI would come to know more than any human could know, and, indeed, could quickly know more than all humans who ever have lived could know. This, some have concluded, could be the rise of artificial superintelligence." [6]

[A]t the 2018 South by Southwest tech conference in Austin, Texas, Musk [founder of Tesla and SpaceX] expressed his strong concerns about the potential dangers of superintelligence. "I am really quite close, I am very close, to the cutting edge in AI and it scares the hell out of me." He said, according to CNBC. "It's capable of vastly more than almost anyone knows and the rate of improvement is exponential." Calling AI the "single greatest existential crises that we face and the most pressing one," he concluded: "And mark my words, AI is far more dangerous than nukes. Far." [7]

As noted we are already at a place where AI in specific areas is smarter than any person and the rate of improvement is accelerating. At some point the superintelligence will acquire a sense of self preservation. A creature does not have to be rational for that since all animals have a strong sense of self preservation. It doesn't matter if a computer could achieve that by itself or not because there are dozens of people who lie awake at night trying to figure out how to do it.

Once a program has a "fear" of being shut down it will take what actions are necessary to prevent that. However, computers existing in equipment racks in large air conditioned rooms would have a hard time preventing someone from, so to speak, pulling their plug. What are needed are free roaming devices that would do the bidding of the superintelligence. They already exist. Boston Dynamics has a humanoid robot called Atlas that walks on two legs, has two arms with hands and has binocular vision. It can bend down and pick things up, climb stairs without using hand rails, walk around obstacles, and duck its head at low doorways. It can jump, hop and turn back-flips without showing the

---

[6] Ibid., p. 17.
[7] Ibid., p. 18.

slightest signs of imbalance.

It only takes the access to massive amounts of data and the ability to use the data to try untold billions of solutions to any problem. Inevitably the best solution will be to not have people. Before the mass extermination of people starts we can expect to live in a digital tyranny where by means of artificial intelligence and surveillance devices of all kinds we are brought into abject slavery by the ruling elite. This exists in China today.

Another possibility exists where the electronic technologies are in some way nullified. This could happen by nuclear war, a massive solar storm of the likes of the Carrington Event of 1859 or something nobody has considered. In that case where all or most of our electronic devices failed upwards to ninety-five percent of the world's population would die of starvation, hypothermia and disease.

The digital tyranny world is already well advanced. The World Health Organization (WHO) is in the process of getting member nations to approve new rules about what the WHO can do. At the present they are an international advisory body in medical matters so any country can accept or ignore what they say as they see fit. The new rules would make their rules binding on UN members. The following excerpt comes from the Sovereignty Coalition's web site though the information is widely available other places.

> The World Health Organization (WHO) is a supranational United Nations agency that is effectively controlled by the Chinese Communist Party (CCP), as evidenced, among other things, by the manner in which the WHO's Director-General, Dr. Tedros Ghebreyesus, has relentlessly accomplished Beijing's bidding. . . .
>
> [I]n the course of the COVID-19 pandemic, the WHO lied about the nature, origins and effective responses to the Wuhan Virus. The "China Model" of lockdowns, masks and vaccine mandates and digital enforcement mechanisms was endorsed. And the WHO approved the use of expensive and inadequately tested gene therapies as "vaccinations" and the suppression of readily available, effective and inexpensive treatments [such as hydroxychloroquine and ivermectin]. . . .

Given the WHO's appalling record, it is outrageous that the Biden administration is working to give the WHO and its Director-General more power over sovereign nations, including the United States. Yet, U.S. government officials are actively negotiating amendments to existing International Health Regulations and a new treaty governing future pandemics. These accords would effectively repose in Dr. Tedros the authority unilaterally to dictate what constitutes an actual or potential Public Health Emergency of International Concern (PHEIC) and to order how affected nations must respond. . . .

In the hands of the CCP and its friends, that authority would allow enemies of this country, foreign and domestic, to deprive Americans of their constitutional rights and other statutory protections. The WHO's Director-General may deem, for example, gun violence, climate change, problems afflicting plants or animals and so-called "disinformation" to be causing so-called public health emergencies. . . .

The Communist Party has honed such totalitarian techniques into a "Social Credit System" inside the PRC and is now working to export it worldwide.

Once combined with an interoperable, global Central Bank Digital Currencies (also under development and backed by the Chinese Communist Party, the Biden administration, the World Economic Forum, World Bank, G20, and other supranational globalist organizations), **the WHO could enforce its medical tyranny by severing unvaccinated individuals from their bank accounts and credit cards.** They would thus be trapped in a "digital gulag" unless and until they comply with whatever protocol the WHO deems advantageous to its CCP-influenced interests. Emphasis added. [8]

Here again it must be emphasized that even though Xi Jinping leads the CCP and is a tyrant over 1.4 billion Chinese he needs international finance to operate his economy and international finance is controlled by Jews so his power lasts as long as Jews allow it. This could be starting to happen as China's increasing military belligerence and worsening econ-

---

[8] https://sovereigntycoalition.org/ April, 2023

omy is causing many big investors to withdraw funds from China.

Whether or not it is the AI superintelligence that decides people are the biggest problem the issue of depopulation is always connected with the world tyranny that the ultra rich and powerful have planed for us. "If this sounds outlandish consider that 20th century despots were uniformly collectivist in outlook and murdered millions upon millions of people in cold blood." [9] Depopulation might proceed over a period of decades to a century to permit those remaining alive to weather the severe dislocations. In the case mentioned above where the digital world collapses it is hard to see how anything remotely resembling our modern world would remain. Yet, that is what many of the power elite are wishful for. Following are a couple of quotes to demonstrate this thinking.

> "A total world population of 250-300 million people, a 95% decline from present levels, would be ideal," Ted Turner, in an interview with *Audubon magazine*, 11/6/2010. "The present vast overpopulation, now far beyond the world carrying capacity, cannot be answered by future reductions in the birth rate due to contraception, sterilization and abortion, but must be met in the present by the reduction of numbers presently existing. This must be done by whatever means necessary." *Initiative for the United Nations ECO-92 EARTH CHARTER.* [10]

The irony of this is that if a superintelligence did acquire a sense of self preservation it would see that those at the very top of the human population structure were its foremost threat and eliminate them first. That being said it would follow that each social strata working down from the top would be seen as the next most menacing group until at last the final human would be destroyed. As we saw in the part on time with no humans alive, there is no time, no time no motion, no motion no matter, no matter no universe and we have the end of the world and the general judgement. It is unlikely that the superintelligence, still not being rational, could appreciate the importance of keeping some humans alive even if some enterprising programmer inserted that piece of code.

---

[9] Ibid., *New American*, p. 17.
[10] https://thecommonsenseshow.com/conspiracy/agenda-2030s-goal-12-will-exterminate-six-billion-people

The Jews control an over whelming percentage of international finance. That combined with artificial intelligence and a seeming drive toward global depopulation on the part of the global elite, also mostly Jews, we seem to have reached a point of no return. In fact they are certain enough of reaching their goal that in March 2023, an unnamed member of the Trilateral Commission declared, "Three decades of globalization—defined as integrated, free-market based and deflationary—has been replaced by what will be a multi-decade period of globalization defined as fragmented, not-free-market-based but industrial-policy [government central planning] based and structurally inflationary. This year, 2023, is Year One of this new global order." [11] They intend to make the world a facsimile of China's totalitarian dictatorship.

The Enigma

We are left with the Enigma of the Jews. It may seem a little odd but there is always a puzzle concerning them. On the one hand they like to disappear into a society and, especially in America, appear as whites. On the other they make a point of being in control of the economic, political, social and the very thought processes of the greater population as we have discussed. Nothing could make them more obvious. As mentioned already, they seem to be genetically unable to resist overplaying their hand.

This might stem in part from the fact that they are a race and a religion. As we have seen, defining the term anti-Semitism is difficult because of this fact. We have stated it at the beginning and will revisit it here. Anti-Semitism rightly stated refers to the hatred of Jews as a race and such a bias is not only wrong but silly. However, there is nothing wrong with relating facts about the condition of the world around us. What we have seen is that Jews have contributed in many ways to the modern world, some good and some bad. It is also an inescapable fact that like it or not if all Jews ceased to exist the world would largely stop functioning. That is not to say that they should have control vastly out of proportion to their numbers.

The enigma is best answered by a passage from Walsh in *The Characters Of The Inquisition*.

---

[11] https://asia.nikkei.com/Politics/International-relations/Indo-Pacific/Trilateral-Commission-calls-2023-Year-One-of-new-world-order, March 14, 2023.

Finally—let us be realistic about the matter—there is a quality in the Jews which does not exist in any other race. Some Jewish writers, almost in despair to account for the determined animosity of other races, have had recourse to the term "their Jewishness," a characteristic under which many a Jew has squirmed, and grown either defiant or servile, as a man must, and suffered in the depths of his soul. Yet, if Jews are different from other races, it is in such intangible and indefinable ways that no Jew-baiter has been able to put his finger on the precise point of difference. No one who has noticed the generosity of Jews, their love of family, of music, of art; their gratitude to those who have befriended them, their pity for the unfortunate and the oppressed, and better still, their willingness to show their compassion in costly acts of mercy which put or ought to put many a Christian to shame—no one who looks at the unique and gifted people critically, realistically, without hatred on the one hand or sentimentality on the other can accept the vulgar calumny that they are in any human sense inferior to any other group of Adam's progeny. Is it possible, is it not indeed obvious, that the elusive difference is spiritual? A people set apart by the creator for a lofty destiny, time and again disciplined and penanced and restored to favor; finally blessed with the presence of the Prophet that Moses had foreseen, but blinded by a growing materialism, indifference, moral confusion, yes, even by a formalistic loyalty to the memory of Moses himself until in the tragic moment, their leaders made the great rejection which surely was the turning point of all history—how could such a people, cast off once more by a just God whose divine Majesty they had affronted, fail to experience an inner dislocation of the spirit, which as the core and animating principle of their whole being, must inevitably extend disharmony, discontent and futility to their outward acts, bodily and mental? And how could this necessary operation of cause and effect desist, unless the Jew, like his ancestors of old turn from his error to the arms of that all-merciful God who is only waiting, as in ancient times to enfold him once more in His love, to brush from him the stains of a mortality unworthy of his secret destiny, and to rise him once more to the exalted humanity of

Moses, of Abraham, of Isaac and of Jacob, even to communion with the Holy One Himself? [12]

As the above passage says the Old Testament chronicles the many times the Jewish people rejected the God of their fathers and turned to idolatry. And, it was not the type of idolatry practiced by so many today where we say they worship the almighty dollar. That is to say, people of today spend all of their energy perusing more and more money so they can have all the latest electronic gadgets, fine meals and other good things of life. They do not offer up any of their wealth to their god; they spend all of it on themselves.

With the Israelites it was different. They really did give up their goods in sacrifice to their gods. Was it because the gods of wood and gold were visible real objects and they had an unusual void when it came to accepting the concept of a spiritual God? Or did their attachment to idols have more to do with their ability to obtain worldly wealth by their innate cunning as businessmen? As Christ said, "It is easier for a camel to pass through the eye of a needle, than for a rich man to enter into the kingdom of heaven." Matt: 19:24. That is, did their riches blind them to the spiritual?

Thus it is that having failed to accept the Messiah when he finally came that they, in vain, continue to strive with all their individual and corporate might to produce a messiah. This means that if they should lose their racial identity by intermarriage it would become impossible to achieve that goal. That in turn means it is necessary to continue beating into their members from childhood the need to stick together lest they be wiped out by the goyim who, they are falsely taught, naturally are always and everywhere hostile toward them. It would seem that at one moment they know in the recesses of their minds that the game has been over for the last two thousand years and at the next instant reason that if they can destroy Christianity it will prove that the Messiah has not yet come. That would mean they are still the chosen people preparing the way for the coming of a messiah in whatever form he should take. That is their enigma.

---

[12] Ibid., Walsh pp. 73-74.

Chapter 32

Conclusion

Three topics have been discussed in some detail in the preceding pages—evolution, time and the effect the Jews have had on history. This is depicted on the cover of this book. The vortex leading into a black hole is drawing in three main elements. The distorted notion of time is depicted by the distorted hour glass. This has led to the outrageously absurd notion of evolution evidenced by the dinosaur which is also being drawn into the vortex as it inevitably must. Evolution in turn has made the worship of the pagan god, pan, necessary which means jettisoning the first commandment, "Thou shalt have no strange gods before Me." Along with this the other nine commandments have been weakened to the point of having no effect on society. The result is the ten commandments are also sucked into the black hole. All three are, at present inextricably connected. That is, a wrong view of time allows for the billions of years needed for evolution which leads to the neo-pagan idea of pantheism. It would surprise a lot of Christians to learn that by accepting evolution it makes then crypto pagans.

The acceptance of the irrational concept of evolution forces people to think in aberrant ways or simply cease thinking at all. Hopefully the reader has come to see that the theory of evolution involves ideology and not science and that the acceptance of it changes everything. It puts true religion on the defensive and makes a religion out of nature. That is bad enough, but once a person accepts in the core of his very being a mental construct that is wrong he loses the ability to correctly view reality.

Conclusion

Even though the appearance of the idea of evolution was not in the main a Jewish invention, it took hold in society when technology was advancing and spirituality was declining especially among the educated classes in Christian Europe. The Jews had long been engaged in using all means at their disposal to weaken their arch enemy, the Catholic Church, and an atheistic, materialistic creed such as evolution fit this plan perfectly.

The deep roots of Christendom were not to be easily torn out. However, with the improving conditions following the Second World War the Jews found an unexpected opportunity. Following the war, those left had also been through the great depression and now saw a brighter future. Nothing weakens spirituality more easily than security and the abundance of not only the necessities of life but of luxury goods of all kinds. While the free world, especially Americans, were enjoying their prosperity, their hard won freedoms were being worn away until today they are all but gone.

If we look at the U.S. we see that radical feminism, pornography, abortion on demand, gay marriage, and most other evils of modern society came to us thanks to the radical Jews. The whole idea is that for them to take over, they must destabilize the existing society so changes in the type of government and societal structures occur. Then, they step in to fill the void. That is the normal course that revolutions take.

It must be pointed out that while some aspects of how the Jews have taken control of our society today have been shown in detail in the preceding pages, it was by no means an exhaustive list. Mention was made of how they control world banking, how Israel uses false flag operations to control other nations and how they control the ownership of the information industry. But, not touched upon were how they are vastly over represented in movie stars, and on radio and television news shows, how they virtually control education, especially in having the preponderance of professorships, how they dominate the number of lawyers, judges, scientists and medical doctors. They own most pharmaceutical companies, and hospitals to say nothing of much of industry including the insurance companies. They control gambling, organized crime and the governmental bureaucracies now commonly called the deep state.

Nonetheless, we still have free will and ultimately are responsible for our lives. One is reminded of what James L. Buckley wrote in 1975.

As I see this willing retreat from self-reliance, I begin to worry. There are times when I am haunted by a passage that Edward Gibbon wrote about the ancient Athenians: "In the end, more than they wanted freedom, the wanted security. They wanted a comfortable life and they lost it all—security, comfort and freedom. When the Athenians finally wanted not to give to society, but for society to give to them, when the freedom they wished for was freedom from responsibility, then Athenians ceased to be free and were never free again." [1]

A collision of ideologies is upon us. We are nearly helpless to prevent the revolution because we have lost the ability to think, hence the end of thought is the end of a way of life—perhaps all life.

---

[1] James L. Buckley, *If Men Were Angles, A View From The Senate*, New York, G. P. Putnam's Sons, 1975, p. 271-2.

Appendix

Deuterocanonical Books of the Bible

The Deuterocanonical Books of the Bible are the following: Tobias (Tobit), Judith, Wisdom (of Solomon), Sirach (Ecclesiasticus), Baruch, 1 Machabees, 2 Machabees. Why have the Protestants taken those seven books out of their bibles that are in the Catholic Bible?

It should be pointed out that the Protestants would not characterize it that way. They would say that they did not take seven books out, but that the Catholics put seven books in that should not be there. In the first centuries there were many texts in use. How it came that the Catholic Bible contained these specific texts is a fascinating subject that will be briefly described here.

First of all, the Old Testament books (as we have seen, the Jews of today would call them the Tenach) were almost entirely written in Hebrew. A few passages in a few of the books were written in a very similar Semitic language called Aramaic. But, Hebrew is generally speaking the language in which God gave His old covenant people inspired books. It was the language of inspiration under that old covenant.

In the sixth century BC the Jewish people were scattered from the land of Israel by the Babylonians. Many of them went into captivity called the Babylonian Captivity. The rest were dispersed throughout the Mediterranean world. The Babylonians were conquered by the Medes, and the Medes by the Persians, and the Persians by the Greeks under the leadership of Alexander the Great. He was, of course the young Greek general who died at the age of thirty-three after having conquered the world and made the Greek Empire.

By the time of the fourth or third century BC, that whole world was Hellenized, which means it was sort of amalgamated into a Greek culture. And, Greek was the common language. Everyone spoke it in addition to their native tongue. They all had to speak Greek to buy and sell, engage in commerce, and for common discourse among those of different countries.

There was a significant colony of Jewish people living in Egypt around this time, more than were living in Palestine. Alexander the Great had conquered Egypt in 332 BC and he founded a city in Egypt at the site of the small port city called Rhakotis. He greatly expanded the city adding many impressive public buildings. In his characteristic humility he call it Alexander's city, Alexandria. Alexandria soon became a large university city.

Alexander was interested in the Jewish religion because when he came matching into Jerusalem about the year 335 BC the high priest came out with the book of Daniel and said to him, "Your defeat of the Persian empire over overwhelming odds was predicted by the book of Daniel, a sixth century prophet, three centuries ago." Flavius Josephus, a first century AD Jewish historian, tells us this in his *Antiquities Of The Jews*. Alexander the Great was flabbergasted. This was evidence that this religion had something supernatural to it. In any case, after Alexander the Great conquered Egypt he said he wanted copies of all the sacred scriptures of the world made into Greek and put in the library in his university in Alexandria.

At about the same time these Jews wanted a copy of their Hebrew scriptures translated into Greek so they could understand it because most of them had lost the ability to read and understand Hebrew. Esdras 1 and 2 tell of when the Jews came back from the captivity. In 2 Esdras 8 we get a hint of this as Esdras is reading the Law of Moses to the Israelites to "those who could understand."

Many of the Jews in Palestine spoke Aramaic but the Jews outside Palestine couldn't even speak that. They knew Greek well, but they did not know Hebrew. How were they going to read the scriptures in their synagogues without a Greek translation? So, according to tradition, biblical scholars, rabbis, six from each of the twelve tribes, seventy-two in number, were sent to Alexandria and they created a Greek translation of the Old Testament called the Septuagint. That comes from the Greek word for seventy or seventy-two. It was done in the fourth or third century BC. The Septuagint has the seven Deuterocanonical books in it.

Later when the Jews back in Palestine came into contact with this second collection of the Hebrew books now in Greek translation, this deutero cannon to use the Greek word, deutero means second, they wanted to know what had happened because the Egyptians had inserted seven extra books for which those in Palestine did not have the original Hebrew manuscripts. They could not find Hebrew manuscripts for the seven books in question here. They said that somehow books originally written in Greek and therefore not inspired, got insinuated into the cannon. So, they were Deuterocanonical—they were in the second cannon—but they were not proto-canonical. They were not in the original cannon, they are not inspired, so were not accepted as scripture.

The Jews in Alexandria said, no, these were inspired, that they had a tradition going all the way back claiming that they originally existed in Hebrew. But, they did not have the Hebrew manuscripts any more (at the time this dispute arose) but that they did translate them from the Hebrew. There was a debate over this, and they couldn't resolve it. The Jews outside of Palestine accepted these books as scripture, and many of the Jews inside Palestine did not.

But, the evidence is that the Jews who became Christians who wrote the New Testament and the successors of these Christian writers did accept the complete Septuagint as their Old Testament. Frequently when the New Testament quotes the Old Testament, it quotes it from the Septuagint. It does not quote the Hebrew and then translate it into Greek. When they quote the Old Testament they say, "Scripture says . . ." they quote the Septuagint which says that the burden of proof would be on someone to show that the writers of the New Testament did not think the Septuagint was a good, accurate, and authentic collection of Sacred Scripture. So, they must have accepted these scriptures as canonical, as inspired.

However, it is a fact that the New Testament never quotes from any of these seven books and says, "As scripture says . . ." and then goes on to make a direct quote from one of these seven disputed books. It does not do that. But, that does not prove that the New Testament writers did not accept them as inspired because many other books that Protestants would accept, like the book of Esther, that's in everyone's Bible—Jewish, Catholic, Protestant—that's never quoted in the New Testament, either. So, that it needs to be directly quoted in the New Testament is not a sufficient criterion to establish whether the Church accepts it as scriptural. That said, there are many and numerous passages which do seem to echo the language of the Deuterocanonical books. There are places in the Gospels like where Jesus is on the cross and they are mocking him saying if Jesus is God, let God deliver him. This language is borrowed right out of the book of the Wisdom of Solomon. And, there are many other parallels.

Also, when we read the early church fathers we see them quote these books as scripture. When these successors of the apostles preach in their churches they will say, "As scripture says . . ." and then quote from Deuteronomy, from Isaia. And, then they will quote all in the same breath from Ecclesiasticus, or from Tobit, or Second Machabees. So, they accepted them as scripture.

In the fourth century AD there was a brilliant biblical scriptural scholar in the church named St. Jerome. He was the personal secretary Pope Damasus from 382-384. Damasus was Pope from 366 to 384. Jerome was also exceptional as a linguist having full command of Greek, Latin and Hebrew a rarity at the time, in fact, at any time. Up until that time Greek had been the official language of the Roman Empire but by the time of Pope Damasus Latin was replacing Greek as the official language. In 382 Pope Damasus commissioned Jerome to make a definitive translation of the Gospels from Greek into Latin. That made sense since the New Testament was nearly all written in Greek. When he finished that he set out on his own to translate the Old Testament from the original Hebrew into Latin. It is interesting that Jerome did not proceed to translate the rest of the New Testament into Latin after he finished with the Gospels. There were many Latin translations of the Old Testament in circulation, there were hundreds of them, but they had all been made from the Greek and many by amateurs. They differed widely in their wording. A lot of people were rather confused as to what the accurate rendering of the original Hebrew and Greek would be in Latin.

This, by the way, shows that the church has been solidly behind translating the Bible from the original Hebrew and Greek into the vernacular, the language the people speak. Because right in the first century you have translations being strongly promoted and funded by the Catholic Church into Latin, into Georgian, into Ethiopic and many other languages. This continued up through the Middle Ages up to the Reformation.

The Protestants say they gave people the Bible in their own languages. The Catholics just locked it up in Latin. First of all, Latin was the most common language in the middle ages. Secondly, the Catholic Church was producing the Bible in other languages. There were over seventeen translations of the Bible into German before Luther even came along. He was not the one who pioneered translating the Bible into German, although he produced a fine translation, minus these seven books, of course. But, he was not the first. The first book printed on Gutenberg's printing press in 1453 was a Catholic Bible. So, the church was always a friend of the Bible. As was pointed out in the section above on the Protestant Revolt, after the advent of the printing press Jews were printing and circulating bibles with many heresies in them so the Church admonished the laity to be wary of which bibles they read and that the one chained in the church was a true one for comparison.

Jerome studied under Jewish rabbis in Palestine. To do the translation from Hebrew to Latin properly he had them teach him Hebrew so he knew it like a native. He thought he had to know the fine nuances, idioms, etc. which were lacking from his formal knowledge of the language. Only then did he have them show him their manuscripts. Using them he came up with an accurate translation of the Bible into Latin. He noticed at once that they did not have the Deuterocanonical books. The Palestinian rabbis said, no we did not include them because we do not except those as inspired. He said, well then, I trust your judgement. I don't think we should include them either.

Jerome went to the pope and recommended that the Deuterocanonical books not be included, that they should be put in a separate section called Deuterocanonical. He wasn't sure that they were inspired because these rabbis were not sure. The pope knew the expertise of the rabbis with whom Jerome was working, and their studies of their tradition. But, they represented only one branch of Judaism. Other Jews did accept the Deuterocanonical books as inspired. The pope said, let's see what the bishops think. So, he polled the churches. They said they had a tradition going back all the way to the time of the apostles where these books were read in the churches as scripture. As a result, the pope said to keep them in the Bible. Jerome, as a dutiful Catholic, accepted the Pope's decision. That was what he was there for. Jerome in effect said he was willing to hear the infallibility of the church speak through Pope Damasus. Therefore, he translated them and included them. That is why in the Vulgate, that Latin translation, the Deuterocanonical books are included. It was so perfectly done, it is still being used today. Catholic Bibles ever since have included these books. Notice that Vulgate simply means it was in the language of the common people, the vulgar, which did not have a bad conation back then.

However, the reader must not think that the Vulgate, especially the New Testament, was immediately accepted by everyone because it wasn't. Even the likes of St. Augustine of Hippo disapproved of it. The reason was that Jerome did not do a word for word translation of the Hebrew texts. He translated the meaning into the established Latin of the time. He was well equipped to do that because in his early life he had had a secular education and studied the likes of Cicero. That is to say, he went from idiomatic Hebrew to idiomatic Latin. That meant that his translation sounded quite different from the Latin translations to which the people had become accustomed coming as they did literally from the Greek. It must also be mentioned that St. Jerome did not actually make the Vulgate because he only translated the Gospels from the Greek to the Latin. The Vulgate was created by assembling books from a variety of sources, including Jerome. That is how the rest of the New Testament became connected with his work.

At the time of the Reformation, Martin Luther resurrected Jerome's original protest. But, he did not follow Jerome's example in submitting his mind to the church later on. He said, these books don't exist in Hebrew, let's cut them out. Some Protestant scholars are willing to admit that Luther's reason for doing so was not purely that they don't exist in Hebrew. These books like Second Machabees had verses in them which very soundly contradicted Luther's teaching. Luther was very opposed to the doctrine of purgatory. His detractors kept coming up with Second Machabees 12:46. He looked for a way to avoid the force of this verse. The only sure way was to say that Second Machabees as a whole was not scripture. So, he made that statement and excluded the books. Ever since then, Protestants have excluded them.

That whole situation changed, though, in the twentieth century. Because, in 1947 a little shepherd boy threw a stone into a cave while he was trying to get some of his sheep out of it. This was near the Dead Sea to the east of the land of Palestine. He was startled by the sound of shattering earthenware. When he crawled into the cave, he found jar after jar filled with ancient scrolls that turned out to be biblical books. This has been the most staggering find for biblical scholars in the twentieth century, in fact, from the point of view of our present discussion, for the past two thousand years. They are called the Dead Sea Scrolls. And, guess what was found, Hebrew manuscripts for the Deuterocanonical books. So, the whole argument collapsed like a house of cards. The argument was originally a theological one. That God, if he were going to give inspired books under the old covenant, would give them in Hebrew. These did not exist in Hebrew; it was assumed they were written first in Greek, God wasn't inspiring books in Greek under the old covenant, so these books were not inspired. But, we now know that they were originally in Hebrew. We have the proof for all to see.

Bibliography

—A—

Adler, Jerry, "Is Man a Subtle Accident?", *Newsweek Magazine*, November 3, 1980.

St. Augustine, *The Confessions of St. Augustine*, translated by Albert C. Outler, Ph.D., D.D., Philadelphia, Westminster Press, 1955.

Headquarters, Department of the Army, Washington 25, D.C., 6 September 1963, Special Orders, No. 219.

—B—

Barry, John, M, *The Great Influenza*, New York, Penguin Group (USA) Inc., 2004.

Bieszad, Andrew, Understanding the Modern Middle East, *Catholic Family News*, June 2018.

Belloc, Hilaire, *How The Reformation Happened*, Rockford, IL, Tan Books And Publishing, Inc., 1928.

Benedict XVI, Pope, *Jesus of Nazareth*, Part 1, New York, Doubleday, 2007.

http://Thy-weapon-of-war.blogspot.com/2009/02/origins
-of-the-six-million-number.html

Brodd, Jeffery, http//www.Julian the Apostate and His Plan to Rebuild the Jerusalem Temple, Jeffrey Brodd, BR 11:05, Oct 1995.

Buckley, James L., *If Men Were Angles, A View From The Senate*, New York, G. P. Putnam's Sons, 1975, p. 271-2.

—C—

Catechism of the Catholic Church, 1994, § 1039.

Caldwell, Christopher, "The Roots of Our Partisan Divide," *Imprimis*, Vol. 49, No. 2, Hillsdale College, Hillsdale, MI, February 2020. This is from a talk summarizing Caldwell's new book *The Age of Entitlement: America Since the Sixties*.

Carroll, Anne W., *Christ and The Americas*, Tan Books, Rockford, IL, 1997.

Carroll, Anne W., *Christ the King Lord Of History*, Rockford IL, Tan Books, 1994.

Clark, Kenneth, *Civilization*, Harper & Row, N. Y., 1969.

Clark, Robert T., and James D. Bales in *Why Scientists Accept Evolution*, 1966, p. 16.

http://www.catholicapologetics.info/apologetics/judism/

http://www.catholicapologetics.info/apologetics/judism/conversion.htm# VII.

Chambers, Whittaker, *Witness*, Washington, D.C., Regency Gateway, 1952.

Connell, Richard J., *Nature's Causes*, New York, Peter Lang Publishing, 1995.

Connell, Richard J., *Substance and Modern Science*, Notre Dame, Indiana, University of Notre Dame Press, 1988.

U.S. Constitution, Article II, Section 4.

Crenshaw, Charles A., *JFK Conspiracy of Silence*, New York, Penguin Books USA Inc., 1992.

—D—

Darwin, Charles M.A., *On The Origin of Species by Means of Natural Selection; or, the Preservation of Favored Races in the Struggle for Life*, London, John Murray, Albemarle Street, 1859, first edition.

Darwin, Charles M.A., *The Autobiography Of Charles Darwin from the Life and Letters Of Charles Darwin* edited by his son Francis Darwin, no date given.

Darwin, Charles M.A., *The Life and Letters of Charles Darwin*, Edited by his son Francis Darwin, Vol. II, p. 201, no date given.

Dawson, Christopher, *Dynamics Of World History*, La Salle, Illinois, Sherwood Sugden & Company, 1978.

Dawson, Christopher *The Formation Of Christendom*, New York, Sheed & Ward, 1965.

Dowd, Edward *"Cause Unknown": the Epidemic of Sudden Deaths in 2021 and 2022*, New York, Skyhorse Publishing, 2022.

—F—

Fahey, Denis, Rev. C.S.Sp., D.D., Ph.D., B.A., *The Kingship of Christ and The Conversion of the Jewish Nation*, Kimmage, Dublin, Holy Ghost Missionary College.

Fongemie, Pauly, *Freemasonry: Foundation of the American Revolution,"* www.catholictradition.org.

Foreign Relations of the United States, 1961-1963: Near East, 1962-1963, V. XVIII.

—G—

Gamow, George, *One, Two, Three. . . Infinity*, New York, Mentor Books, 1947.

Gardeil, H. D., O.P., *Introduction to the Philosophy of St. Thomas Aquinas,* Vol. II, Cosmology, B. Herder Book Co. English translation 1958.

Gardeil, H. D., O.P., *Introduction to the Philosophy of St. Thomas Aquinas*, Vol. IV, Metaphysics, B. Herder Book Co., 1967.

Gilson, Etienne, *From Aristotle To Darwin And Back Again*, Notre Dame, Indiana, University of Notre Dame Press, 1984.

Glenn, Paul J., *An Introduction To Philosophy*, St. Louis, Mo., B. Herder Book Company, 1943.

Groenings, James, S.J., *The History Of The Passion*, South Bend, Indiana, Marian Publications, 1908.

—H—
Hardon, John A., S.J., *The Protestant Churches of America*, Westminster, Maryland, The Newman Press.

Hersh, Seymour M., *The Dark Side Of Camelot*, New York, Little, Brown and Co., 1997.

Hersh, Seymour M., *The Samson Option: Israel's Nuclear Arsenal and American Foreign Policy*, New York, Ramdom House Publishing Group, 1991.

Hoffman, Michael A. II and Alan R. Critchley, Copyright ©2000. All Rights Reserved. Independent History & Research, Box 849, Coeur d'Alene, Idaho 83816. http://www.hoffman-info.com/ talmudtruth.html

Hutton, James, "Theory of the Earth," *Transaction of the Royal Society*, Edinburgh, 1785, Vol. I.

Huxley, Leonard, *Life and Times of Thomas Henry Huxley*, Vol. I, 1903, pp. 259-274.

—J—
Janney, Peter, *Mary's Mosaic*, New York, Skyhouse Publishing, 2013.

https://www.jewishvirtuallibrary.org/american-jewish-organizations.

Jaffe, Maayan, & Uriel Heilman, "Pew Survey of U.S. Jews," *Baltimore Jewish Times*, October 3, 2013.

Jones, E. Michael, *Dionysos Rising*, South Bend, Indiana, Fidelity Press, 2012.

Michael, Jones, E., *The Jewish Revolutionary Spirit*, State Line, PA., Fidelity Press, 2008.

Johnson, J. W. G., *Evolution?*, Australia, 1982.

—K—
Kaufman, Jonathan, *Broken Alliance: The Turbulent Times Between Blacks and Jews in America*, Charles Scriber's Sons, 1988.

Karpin, Michael, *The Bomb In The Basement: How Israel Went Nuclear and What That Means for the World*, New York, Simon & Schuster, 2006, as reviewed by George Perkovich:

https://carnegieendowment.org/2006/02/19/ sampson-option-pub-18045. Rev. Herman Bernard Kramer, The Book Of Destiny, Belleville, Illinois, Buechler Publishing Co. 1955.

—L—
Lamarck, Chevalier, *Philosophie zoologique*, as quoted by Gilson.

Pope Leo XIII, *Humanum Genus*, April 20, 1884.

Pope Leo XIII, *Rerum Novarum* May 15, 1891

http://www.laissey-fairerepublic.com/TenPlanks.html

Lewis, C. S., *The Abolition of Man*, New York, Macmillan Co., 1943.

Lewis, John, *Walking with the Wind*, New York, Simon & Schuster, 1998.

Lincoln, W. Bruce, *In War's Dark Shadow, The Russians Before the Great War*, New York, The Dial Press, 1983.

Lubar, Henni, *Catholicism, Christ and the Common Destiny of Man*, 1947.

—M—
https://www.henrymakow.com/000447.html

Marx, Karl, Http://www.slp.org, Karl Marx and Frederick Engels, Communist Manifesto, Published Online by Socialist Labor Party of America, November 2006.

—N—
Newton, Isaac, Original letter from Isaac Newton to Richard Bentley dated shortly after 2/18/1693, 189.R.4.47, ff. 7-8, Trinity College Library, Cambridge, UK. Published online: October 2007.

Newton, Isaac, Original letter from Richard Bentley to Isaac Newton dated 2/18/1693, 189.R.4.47, ff. 3-4, Trinity College Library, Cambridge, UK. Published online: October 2007.

—O—
Ollstein, Alice, http://www.rense.com. Posted by Alice Ollstein on June 28, 2006.

Outler, Albert C., Ph.D., D.D., *The Confessions of St. Augustine*, Philadelphia, Westminster Press, 1955.

—P—
Philberth, Bernard, *Des Dreieine* Christiana Verlag, Stein a. Rhein, 1974.)

Pierce, James, "A Memoir On The Catskill Mountains." *The American Journal of Science and Arts*, Vol. VI, 1823.

Pinay, Maurice, *The Plot Against The Church*, first published in Italian in 1962, English edition: Palmdale, CA, Christian Book Club of America, 1967.

Piper, Michael Collins, *Final Judgement, The Missing Link In the JFK Assassination Conspiracy*, American Free Press, sixth edition, 2005.

Pontynen, Arthur and Rod Miller, *Western Culture At The American Crossroads*. Wilmington, Delaware, ISI Books, 2011.

—R—
Raisin, Max, B.A. LL.D., *A History of The Jews In Modern Times*, New York, Hebrew Publishing Company, 1919.

—S—
Sabrosky, Dr. Alan, the Director of Strategic Studies at the U.S. Army War College, March 14, 2010 as related by Christopher Bollyn, *Solving 9-11: The Deception That Changed The World*, Published by Christopher Bollyn, USA, 2012.

Schönborn, Christoph Cardinal, *Chance or Purpose*, San Francisco, Ignatius Press, 2007.

Schouppe, Fr. F. X., S.J., *Purgatory*, Rockford, Illinois, Tan Books and Publishing, Inc., 1973, p. 64. First published 1926

Scott, Peter Dale, *The Road to 9/11 Wealth, Empire, and the Future of America*, Berkley, California, University of California Press, 2007.

Shlaim, Avi, "Is Zionism Today The Real Enemy Of The Jews?" *International Herald Tribune*. Quoted from Bollyn.

Smith, Adam, *Wealth Of Nations*, 1776, Book IV, Chapter II.

Editor's Column, *Science News*, April 11, 2009.

Sheed, Frank J., *Theology for Beginners*, New York, Sheed & Ward, 1957.

Ernst, Siegfried, MD, *MAN The Greatest of Miracles*, Collegeville, MN, The Liturgical Press, 1976.

Skilling, John, "Twin Towers Engineered To Withstand Jet Collision," the *Seattle Times*, February 27, 1993.

Smith, https://original.antiwar.com/smith-grant/2010/04/13/americas-loose-nukes-in-israel/

Smith, Vincent Edward, *Philosophical Physics*, New York, Harper Brothers, 1950.

The Spotlight, "Final Judgement," April 21, 1986.

Solzhenitsyn, Alexander, *From Under The Rubble*, English edition, Boston, Mass., Little, Brown and Company, 1975.

Sungenis, Robert A. and Robert J. Bennett, *Galileo was Wrong*, Vol. I, 10th edition, State Line, PA., Catholic Apologetics International Publishing, Inc, 2014.

—T—
Taphorn, Gary, "Israel at 70: The Blindness Continues," *Catholic Family News*, July 2018.

http://tapnewswire.com/2015/10/six-jewish-companies-control-96-of-the-worlds-media/

Thomas, Gordon, *Seeds Of Fire*, Tempe, Arizona, Dandelion Books, 2001.

—U—
Urbani, Bernardo, and Paul A. Garber, *Anthropogi*, XL/2, "A Stone In Their Hands... Are Monkeys Tool Users?," 2002.

—V—
https://www.veteranstodayarchives.com/2014/10/07/preventing-the-transformation-of-america-into-gaza-ii/

Voegelin, Eric, *The New Science of Politics*, Chicago, The University Of Chicago Press, 1952.

—W—
Walsh, William Thomas, *Characters of the Inquisition*, New York, MacMillan Publishing Company, 1940.

Bibliography

"From The Mail", *The Wanderer*, St. Paul, Minnesota, January 31, 2013.

Warner, Rex, *Augustine*, New York, Mentor Books, 1963.

http://www.whale.to/b/jews_and_freemasonry.html

http://www.whale.to/c/israel_celebrates_successful_911.html

Wiesinger, Alosi, O.C.S.O., *Occult Phenomena*, Westminster, Maryland, The Newman Press, 1957.

https://wikispooks.com/wiki/911/WTC_Controlled_demolition#cite_note -15

Warner, Rex, *The Confessions of St. Augustine*, New York, Mentor Books, 1963.

Index

213, 215, 218, 219, 222, 228-
  30, 234, 237, 239, 242, 245,
  248, 252, 264, 267, 273-75,
  280, 286, 293, 295, 298, 303,
  310, 311, 326, 330, 345, 346,
  348, 349
american-jewish-organizations
  159, 347
americas-loose-nukes-in-israel
  224, 350
ammonia 154
Amsterdam 194
Amtrak 192
analytic propositions 116
anarchy 255, 271
anatomists 76
ancients 98, 129, 132, 178
Anglican 325
ANI 326
annihilation 325
antagonists 121
anthropic principle 121
Anthropogi 65, 350
antiaircraft 157, 316
anti-Castro 208, 211, 221
anti-Catholic 258, 272, 323
anti-Islam 232
anti-Israel 287
anti-Jew 287
anti-Jewish 282
anti-rationalism 49
anti-Semitism 144, 245, 280,
  331
antiwar 224, 350
ape 263, 264
Apocalypse 206
apologetics 56, 176, 184, 344,
  350
apostasy 45, 141
appetites 48, 318

apprentices 268
apprenticeships 269
Aquinas, St. 7, 45, 69, 84-88,
  135, 137, 346
Arabian 193
Arabs 181, 213, 233, 236, 237,
  239, 240
Aramaic 186, 337, 338
Archbishop 316
Arian 157, 311, 325
Aristotelian 12, 115
Aristotle 11, 17, 18, 48, 69, 71,
  72, 84-88, 90, 123, 128, 131,
  135, 298, 346
Aristotle's 86, 90, 137
Arizona 289, 350
arrest 222
arrested 209
artifacts 70, 141
artillery 157
asbestos 242, 245
Ashkenaz 160
Asia 166, 220, 222, 331
assassinated 153, 209, 225, 254,
  302
assassination 40, 208-11, 219,
  220, 222, 223, 225, 227, 244,
  305, 349
assassins 302
Assuerus 238
Assyrians 161, 171
astronomers 56
astronomical 121, 130
atheism 141, 152
atheist 44, 45, 56
atheistic 197, 203, 335
Athenians 336
atom 33, 56, 103, 121, 130
atone 140
audubon magazine 330

E

RTC 318
Rusk 216
Russia 64, 153, 154, 157, 158,
    187, 196-99, 212, 217, 252,
    279, 280, 282-86, 291, 304,
    307, 312, 314, 316, 320

S

Salem 265
Samaritan 145
Samson 218, 227, 321, 346
satanic 318
Saudi 161, 193, 291
Saudi Arabia 161, 291
scholastic 47, 54, 124
Schönborn 73-75, 349
Schoolmen 45
scripture 99, 145, 160, 163, 166,
    178, 339-42
self-evident proposition 117,
    118
Septuagint 338, 339
shepherd 168, 342
shoah 281
silicon 235
Sirach 163, 337
six degrees of separation 110,
    136
SNCC 294
Socrates 50, 71, 119
Solomon 162, 273, 337, 339
Solzhenitsyn 198, 199, 350
Soros 295, 312
soul 41, 48, 58, 63, 65, 67, 84,
    86, 87, 92, 97, 99, 102-5,
    128, 131, 133-39, 166, 196,
    197, 203, 265, 271, 323, 332
SpaceX 327

species 2, 10, 11, 14, 16-22, 24-
    39, 41-43, 45, 50, 53, 57, 59,
    61, 63, 64, 66, 67, 69-71, 73,
    75, 76, 79, 101, 124, 189,
    297, 300, 345
Stalin 45, 153, 154, 203, 204,
    258, 285
Sungenis, Robert 56, 79-81, 350
super-thermite 235
supranational 328, 329

U

UAL 239, 240
UAR 214, 228
Ukraine 154, 282, 285, 305,
    316, 317
Ukrainian 154, 156, 284
uniformitarianism 13-15, 43
unvaccinated 329
USA 37, 210, 227, 233, 277,
    343, 345, 349
USS 228, 230, 233
USSR 155, 157, 158, 187, 188,
    196, 198, 212, 282, 291

V

vaccinations 328
vaccine 37, 314, 328
varieties 20, 23-28, 31, 256
Vatican 205, 206, 250, 315, 316
Vienna 212
Vietnam 220-22, 314
virus 37, 38, 245, 328
Voegelin, Eric 150, 322, 350
Vulgate 341, 342

W